Low-Dimensional Magnetism

Low-Dimensional Magnetism

A.N. Vasiliev

O.S. Volkova

E.A. Zvereva

M.M. Markina

CISP

CRC Press
Taylor & Francis Group
Boca Raton London New York

CRC Press is an imprint of the
Taylor & Francis Group, an **informa** business

Translated from Russian by V.E. Riecansky

CRC Press
Taylor & Francis Group
6000 Broken Sound Parkway NW, Suite 300
Boca Raton, FL 33487-2742

First issued in paperback 2021

© 2019 by CISP
CRC Press is an imprint of Taylor & Francis Group, an Informa business

No claim to original U.S. Government works

ISBN 13: 978-1-03-223900-2 (pbk)
ISBN 13: 978-0-367-25535-0 (hbk)

Publisher's Note
The publisher has gone to great lengths to ensure the quality of this reprint but points out that some imperfections in the original copies may be apparent.

Library of Congress Cataloging-in-Publication Data

Names: Vasiliev, A. N. (Alexander N.), author.
Title: Low-dimensional magnetism / A.N. Vasiliev [and three others].
Description: Boca Raton : CRC Press, Taylor & Francis Group, 2019. | Includes
    bibliographical references and index.
Identifiers: LCCN 2019021450 | ISBN 9780367255350 (hardback : alk. paper)
Subjects: LCSH: Magnetism. | Magnets.
Classification: LCC QC753.2 .L69 2019 | DDC 620.1/1297--dc23
LC record available at https://lccn.loc.gov/2019021450

Visit the Taylor & Francis Web site at
http://www.taylorandfrancis.com

and the CRC Press Web site at
http://www.crcpress.com

Contents

Contents

Introduction

The key to understanding the fundamental properties of matter is at low temperatures. Under the conditions where thermal oscillations do not conceal low-energy interactions, a field of quantum cooperative phenomena that have no analogs in classical physics opens up. It is these phenomena – superconductivity, exotic magnetism, spin and charge density waves, Bose–Einstein condensation – that are relevant in the physics of the condensed state. Most clearly, these effects are manifested in compounds where magnetoactive ions form objects or clusters of reduced dimensionality, such as dimers, chains, ladders and layers. The work on the study of the physical properties of new magnetic materials is aimed at understanding the most common effects and interactions that form the quantum ground states of matter.

Quantum cooperative phenomena form a special branch of condensed matter physics. It is the quantum aspects in the behaviour of matter 'quantum entanglement', spin-polarized transport, exotic superconductivity – that are at the base or are supposed to be used in the most advanced technologies. Magnetism and superconductivity, considered antipodes for a long time, reveal common features in objects that were previously outside the field of view of theorists and experimenters. Actually, the detection of high-temperature superconductivity in complex oxides of transition metals, which initially are antiferromagnetic insulators, completely changed the vector of the development of solid state physics. Interest began to attract the so-called 'new magnetic substances', i.e., substances with a reduced dimensionality of the magnetic subsystem and the frustration of the exchange interaction. It became clear that in some of these systems the ground state of matter is the spin liquid, and the properties of this state and its elementary excitations are close to those of the electron liquid in superconductors.

As a result of the investigation of the first high-temperature superconductors – copper-based complex oxides – it was found that along with the interaction of the conduction electrons with the lattice, an important part in the formation of the superconducting state belongs to the magnetic subsystem. New high-temperature superconductors – pnictides and iron chalcogenides – also exhibit unusual magnetic properties at high temperatures. This magnetism, unlike the 'classical' magnetism of iron, has a number of principal features, which explains its modern classification as 'new magnetism'. Low-dimensional magnetism is most clearly manifested in frustrated systems, when the formation of the long-range magnetic order is difficult or impossible. In this case, the scale of the exchange magnetic interaction can turn out to be large not only at nitrogen but also at room temperature. In such a situation, magnetism and superconductivity not only compete but also 'support' each other. It is important to note that studies in the field of low-dimensional magnetism are aimed not only at increasing the functional characteristics of magnetic materials, but also form new directions in the physics of condensed matter. First of all, this is the physics of spin liquids, non-collinear and exotic magnetic structures, topological insulators, multiferroelectricity, and quantum superposition of states.

The existing problems in the physics of low-dimensional magnetism are associated with the search for and improvement of the functional parameters of new magnetic compounds, bringing their characteristics in line with the requirements of innovative technologies. To achieve the stated goal, specific tasks are being solved worldwide in parallel to establish the dominant mechanisms of magnetic interaction, to determine the parameters of the exchange interaction in new magnetic materials. As a result of a comprehensive study of these materials, priority data were obtained on the main mechanism of the ground state formation, phase diagrams were constructed and the characteristics of the magnetic subsystem were determined in the formation of the long-range magnetic order. The obtained data stimulated the development of theoretical ideas about the structure of matter.

In magnetic systems where one or several directions lack or have an infinitely small exchange magnetic interaction, the description of physical phenomena is possible only in the language of quantum mechanics. In the simplest case, such a system is two localized magnetic moments. To describe the exchange magnetic interaction in the dimer, the Hamiltonian proposed by Heisenberg in 1928 [1] and the total spin operators $\hat{S}_1$ and $\hat{S}_2$ for centres 1 and 2, formulated in the work of Dirac and van Vleck in the 30s, were used [2]:

$$\hat{H}^{\text{Heisenberg}} = -J\hat{S}_1\hat{S}_2,$$

where J is the exchange interaction constant.

An increase in the number of localized magnetic centres can lead to models of clusters, chains, ladders, two-dimensional layers. In many-particle systems, the Heisenberg Hamiltonian takes into account only the interactions between nearest neighbors $\langle i,j \rangle$:

$$\hat{H}^{\text{Heisenberg}} = -\sum_{\langle i,j \rangle} J_{ij}\hat{S}_i\hat{S}_j.$$

To date, a number of theoretical models have been developed, taking into account the anisotropic terms of the exchange interaction and interactions with the next neighbours, which often leads to different solutions for the quantum ground state of low-dimensional systems. Of greatest interest are situations where it is possible to find experimental confirmation of predicted phenomena in real objects.

Magnetism of compounds with reduced dimensionality of the spin subsystem – zero-dimensional 0D, one-dimensional 1D or two-dimensional 2D – is one of the most interesting and rapidly developing areas in the physics of condensed matter. In low-dimensional magnets, the quantum essence of matter manifests itself most clearly, and it becomes possible to observe a variety of quantum cooperative effects. In 1D, and even more so in the 2D situation, the influence of anisotropy and frustration increases significantly, which considerably complicates the mechanisms of achieving the quantum ground state and enriches the phase diversity on the magnetic phase diagram.

A large number of new fascinating phenomena are observed in low-dimensional magnets, including cascades of spin-reorientation transitions, magnetization plateaux and instabilities induced by the magnetic field of the spin subsystem. Quasi-1D and quasi-2D frustrated systems are often ordered into non-collinear, incommensurable and canted magnetic structures. Such spin states are often devoid of the inversion centre and can have a finite ferroelectric polarization.

Magnetic clusters

1.1. Dimers

For an isolated antiferromagnetic dimer composed of spins $S = 1/2$, the energy spectrum contains the spin gap Δ separating the states with the values of the total spin $S = 0$ and $S = 1$. In the external magnetic field B, due to the Zeeman effect, the levels corresponding to the spin 1 split, as shown in Fig. 1.1.

For such a system, the temperature dependence of the magnetic susceptibility of a mole of ions can be written as:

$$\chi = \frac{g^2 \mu_B^2}{k_B T}\left(3 + \exp\left(\frac{\Delta}{k_B T}\right)\right)^{-1}.\qquad(1.1)$$

In this notation, the gap size and the integral of the exchange magnetic interaction coincide ($\Delta = J$), k_B is the Boltzmann constant.

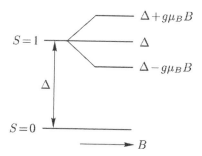

Fig. 1.1. Schematic representation of the energy diagram of the dimer $S = 1/2$ in an external magnetic field B. Δ – energy gap, μ_B – Bohr magneton, g – g-factor.

The $\chi(T)$ dependence shows a maximum whose temperature is determined by the condition $\partial \chi / \partial T = 0$ and is $T \approx 0.625 J/k_B$. In the high-temperature region, the susceptibility obeys the Curie–Weiss law. At low temperatures the susceptibility tends exponentially to zero.

For an isolated antiferromagnetic dimer composed of spins $S = 1/2$, the standard thermodynamic approach makes it possible to obtain a formula for the temperature dependence of the magnetic contribution to the specific heat:

$$C = \frac{3}{2} R \left(\frac{\Delta}{k_B T} \right)^2 \exp \left(-\frac{\Delta}{k_B T} \right) \Big/ \left(1 + 3 \exp \left(-\frac{\Delta}{k_B T} \right) \right)^2. \tag{1.2}$$

At low temperatures, the magnetic contribution to the specific heat is described by the formula:

$$C \propto R \left(\frac{\Delta}{k_B T} \right)^2 \exp \left(-\frac{\Delta}{k_B T} \right). \tag{1.3}$$

With increasing temperature, the specific heat of the molar ions passes through a maximum and at high temperatures is inversely proportional to T^2:

$$C \propto R \left(\frac{\Delta}{2 k_B T} \right)^2. \tag{1.4}$$

In the crystalline structure of real objects, as a rule, there are chemical bonds between the individual dimers. This makes possible the exchange magnetic interaction between them. Such exchange interactions somewhat complicate the general picture of the quantum ground state. Establishing a true picture of exchange interactions is an independent task. To identify this picture, more accurate processing of magnetic data, theoretical calculations from the first principles, inelastic neutron scattering, nuclear magnetic resonance, and electron paramagnetic resonance are used. In some cases, the exchange outside the dimer structural unit dominates.

To describe the temperature dependence of the magnetic susceptibility of interacting dimers, formula (1.1) is applied, modified in the mean-field theory [3]:

$$\chi = \frac{N_A g^2 \mu_B^2}{k_B T}\left(3 + \exp\left(\frac{J}{k_B T}\right) + \frac{J'}{k_B T}\right)^{-1}, \qquad (1.5)$$

where N_A is the Avogadro number, J corresponds to the intra-dimer interaction, and J' corresponds to the overall exchange magnetic interaction with n magnetic centres in neighboring dimers: $J' = \sum_i n_i J_i'$.

1.1.1. Spin gap in cesium divanadate

As will be shown below, interdimer interactions can lead to the formation of more complex magnetic objects, such as an alternating chain or a two-dimensional plane. Their thermodynamic properties will be considered in the corresponding sections. One of the most known spin-dimer compounds based on vanadium is cesium divanadate CsV_2O_5, which has a monoclinic structure (space group $P2_1/c$), which is shown in the left panel of Fig. 1.2. Here, vanadium ions are present in magnetic V^{4+} and non-magnetic V^{5+} states. As shown in the right panel of Fig. 1.2, a broad maximum near $T_{max} \sim 100$ K is present on the temperature dependence of the magnetic susceptibility, accompanied by a drop in the magnetic susceptibility to practically zero and some increase at the lowest temperatures. The low-temperature rise of the $\chi(T)$ dependence is attributed to the presence of paramagnetic impurity centres, whose content does not exceed one percent. The processing of the temperature dependence of the magnetic susceptibility of the matrix obtained after subtracting the impurity contribution made it possible to estimate the value of the

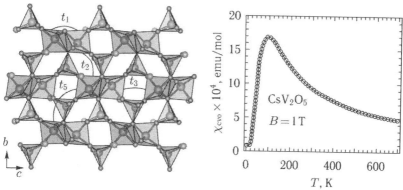

Fig. 1.2. The left panel: the crystal structure of CsV_2O_5 and the scheme of magnetic interactions. Cs^+ ions are shown by isolated spheres, V^{4+} ions are in octahedra, and V^{5+} – ions in tetrahedra. The right panel: the temperature dependence of the magnetic susceptibility CsV_2O_5. Solid line – fit according to the model of interacting dimers [5].

parameter of the antiferromagnetic exchange interaction in the dimer as $J = 146$ K. The first-principle calculations of exchange magnetic interactions in the system performed in [4] showed the presence of a noticeable interdimensional interaction along the way t_3, as shown in the left pane of Fig. 1.2. Thus, to explain the properties of CsV_2O_5, a model of an alternating chain with exchange interactions in the dimer $J = 260$ K and between the dimers $J' = 30$ K.

1.1.2. Mineral urusovite

Mineral urusovite $CuAl(AsO_4)O$ has a monoclinic structure, the space group $P2_1/c$ with four formula units per unit cell [6]. The main structural unit is a distorted CuO_5 pyramid. Two adjacent pyramids are united by a common edge in the basal plane, forming a dimer. Dimers are connected through a common vertex, forming a honeycomb-type structure in the bc plane (the left panel in Fig. 1.3). The 'honeycomb' layers are connected along the a axis through the AsO_4 and AlO_4 tetrahedra (the right panel of Fig. 1.3).

The temperature dependence of the magnetic susceptibility in $CuAl(AsO_4)O$, measured in a field $B = 1$ T in the temperature range 2–1000 K, is shown in Fig. 1.4. It shows the predominant contribution of impurities/defects at low temperatures and the non0monotonic dependence of the signal with increasing temperature. In order to emphasize the contribution of the matrix, measurements of the magnetic moment M were also carried out in a magnetic field $B = 9$ T in the temperature interval 2–400 K, as shown in the inset to Fig. 1.4. At a temperature $T_{max} \sim 215$ K, the magnetization passes through a wide maximum and decreases with increasing temperature.

It should be noted that the authors of this book use the terms 'magnetic susceptibility' and 'reduced magnetization' for the same

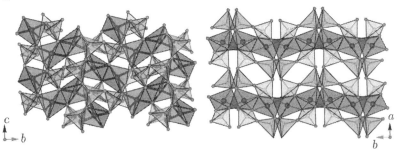

Fig. 1.3. Crystal structure of urusovite $CuAl(AsO_4)O$ in the polyhedral representation. Copper is in the of edge-shared pyramids of CuO_5, arsenic and aluminum – in the tetrahedra of AsO_4 and AlO_4 [6].

expression *M/B*. This is due to the use of modern research techniques, such as the SQUID magnetometer, where the measured value is the magnetization (magnetic moment) *M*. The use of the term 'magnetic susceptibility' is valid for weak magnetic fields (less than 1 T), where *M/B* is close to the definition of susceptibility $\chi = dM/dB$ ($dB \rightarrow 0$), since the dependence *M(B)* is linear for paramagnets and antiferromagnets in this field region. In the range of higher fields, the term 'reduced magnetization' is more correctly used for the *M/B* ratio, since the non-linear dependence of *M(B)* can be observed with increasing field.

The temperature dependence of the magnetic susceptibility $\chi(T)$, measured over a wide range of temperatures, was approximated by the formula of non-interacting dimers (1.1) with the addition of a temperature-independent term and the contribution of impurities/defects described by the Curie law. The dependence obtained as a result of the approximation is shown in Fig. 1.4 by a solid line. The temperature-independent term $\chi_0 = -8 \cdot 10^{-5}$ emu/mol is the sum of the Pascal diamagnetic constants of each ion in the chemical formula $CuAl(AsO_4)O$ and the paramagnetic van Vleck contribution due to the splitting of the *d*-shell of the Cu^{2+} ion in the crystal field. The value of the spin gap Δ, equal to the exchange value in dimers J_1, is 350 K. At $T^* = 800$ K, chemical degradation of the sample takes place and the results are not reproduced in repeated measurements.

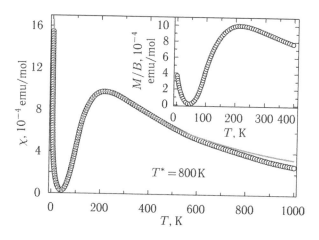

Fig. 1.4. Magnetic susceptibility of $CuAl(AsO_4)O$. The line shows the approximation in the model of isolated dimers. The inset shows the dependence of the reduced magnetization *M/B* in the field *B* = 9 T [6].

The temperature dependence of the specific heat of $CuAl(AsO_4)O$ is shown in Fig. 1.5. The value of the spin gap, determined from measurements of the magnetic susceptibility, was used to calculate the temperature dependence of the magnetic specific heat in accordance with formula (1.2). The subtraction of the magnetic component from the total specific heat yields the lattice contribution of C_{lat}, shown in Fig. 1.5 a dotted line. The lattice contribution is well described by three Einstein modes with energies $E_1 = 112$ K, $E_2 = 287$ K, and $E_3 = 763$ K. These modes can correspond to three basic structural units in $CuAl(AsO_4)O$, i.e., CuO_5 pyramids, AsO_4 and AlO_4 tetrahedra.

At present, experimental studies of new materials are supported, as a rule, by analytical or first-principles calculations. Of decisive importance for these calculations are the interatomic distances and angles, under which the orbitals of ligands and transition metals overlap. Figure 1.6 shows a fragment of the urusovite structure with some Cu–O–Cu angles and Cu–O distances at a temperature of 300 K. As the temperature rises, the thermal expansion of the crystal lattice takes place, and these parameters change. The splitting of the $3d$-shell of Cu^{2+} ions in the pyramidal oxygen environment is shown in Fig. 1.7. The magnetoactive orbital is $3d_{x^2-y^2}$. Table 1.1 shows the interatomic distances and angles for temperatures of 300 and 1000 K. Based on these data, the integrals of the exchange interaction were calculated. At $T = 300$ K they are equal: intradimer exchange $J_1 = 350$ K, interdimer exchange $J_2 = 0.6$ K, interaction between layers $J_3 = 8.3$ K; at $T = 1000$ K, these parameters are $J_1 = 535$ K,

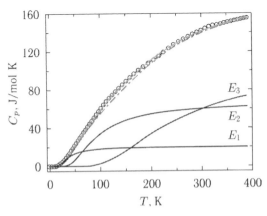

Fig. 1.5. Specific heat of $CuAl(AsO_4)O$. The solid lines show Einstein modes with energies $E_1 = 112$ K, $E_2 = 287$ K, $E_3 = 763$ K [6].

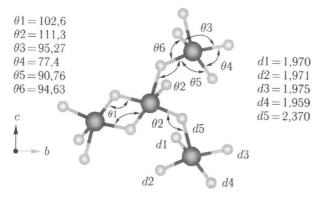

$\theta 1 = 102,6$
$\theta 2 = 111,3$
$\theta 3 = 95,27$
$\theta 4 = 77,4$
$\theta 5 = 90,76$
$\theta 6 = 94,63$

$d1 = 1,970$
$d2 = 1,971$
$d3 = 1,975$
$d4 = 1,959$
$d5 = 2,370$

Fig. 1.6. A fragment of the crystal structure of $CuAl(AsO_4)O$, where the main angles (°) and interatomic distances (Å) are indicated [6].

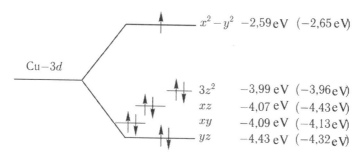

Cu–3d

$x^2 - y^2$	$-2,59\,eV$ $(-2,65\,eV)$
$3z^2$	$-3,99\,eV$ $(-3,96\,eV)$
xz	$-4,07\,eV$ $(-4,43\,eV)$
xy	$-4,09\,eV$ $(-4,13\,eV)$
yz	$-4,43\,eV$ $(-4,32\,eV)$

Fig. 1.7. The energy levels of the 3d orbitals of the Cu^{2+} ion at 300 K (in parentheses are given the values at 1000 K) [6].

$J_2 = 5$ K, $J_3 = 5$ K. The ways of exchange interaction through O^{2-} ions are shown in Fig. 1.8.

The temperature interval of the experimental study of the magnetic susceptibility of uraniumite $CuAl(AsO_4)O$ is much wider than in the usual laboratory practice. Measurements to room temperature, as a rule, do not require consideration of the effect of thermal expansion. With increasing temperature, changes in the interatomic angles and distances can lead to serious corrections in the parameters of the exchange interaction. In addition, the splitting of the electronic d-shell of a transition metal can vary with temperature. As a result, the g-factor and van Vleck's contribution can change. If the g-factor estimate requires information on the scale of the spin-orbit interaction, the contribution of van Vleck can be estimated from the splitting of the d-shell of copper at different temperatures.

Table 1.1. Parameters of the crystal lattice $CuAl(AsO_4)O$

Cu–O–Cu angles, °	$T = 300$ K	$T = 1000$ K	Distances, Å	$T = 300$ K	$T = 1000$ K
θ_1	102.36	105.38	d_1	1.97	1.956
θ_2	111.32	112.62	d_2	1.967	1.955
θ_3	101.07	101.67	d_3	1.97	1.962
θ_4	77.64	74.62	d_4	1.96	1.967
θ_5	90.48	91.68	d_5	2.37	2.306
θ_6	94.88	95.87			

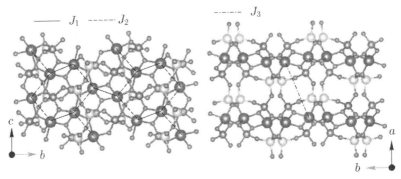

Fig. 1.8. Ways of exchange interaction in $CuAl(AsO_4)O$: intra-dimeric J_1, inter-dimensional J_2 and interlayer J_3 [6].

The discrepancy between the experimental data and the approximation is about 10% at $T^* = 800$ K and 25% at 1000 K. Various chemical and physical factors are responsible for this discrepancy.

Van Vleck's contribution to the magnetic susceptibility can be written as

$$\chi_{vV} = \frac{4N_A\mu_B^2}{\Lambda}, \qquad (1.6)$$

where $\Lambda = |10Dq|$ is the splitting of the d-shell by energy in the crystal field. Since there is a redistribution of the energy levels with a change in temperature, as shown in Fig. 1.7, the total splitting of the copper d-shell changes somewhat, is in the decrease of the paramagnetic contribution. This contribution amounts to $\chi_{vV} = 7.47 \cdot 10^{-5}$ emu/mol at 300 K and $\chi_{vV} = 7.26 \cdot 10^{-5}$ emu/mol at 1000 K. The difference in 3% does not allow to explain the discrepancy between the experimental data and the approximation.

Another factor that affects the absolute magnitude of the magnetic susceptibility χ is the presence of interdimer interactions. Approximation of experimental data by an expression with a large number of independent parameters is unproductive and often ends with a fitting with different sets of parameters.

The intraplanar J_2 and interplanar J_3 exchange interactions between dimers can be taken into account using the formula (1.5), where the total interdimer interaction $J' = 2J_2 + J_3$ is 9.5 K at room temperature and $J' \sim 15$ K at 1000 K. This factor reduces magnetic susceptibility at high temperatures of less than 1%.

The most important factor responsible for the discrepancy between experiment and theory appeared to be an increase in the intradimer exchange interaction of J_1 from 350 K at room temperature to 535 K at 1000 K. The result of this single factor is a 10% decrease in the magnetic susceptibility. The discrepancy between the curves above $T^* = 800$ K can be attributed to the thermal decomposition of the sample.

The long-range magnetic order is not established in the urusovite up to the lowest temperature attained during the experiments of 2 K. Instead, a state with weakly interacting dimers on a lattice of the honeycomb-type is realized. Thus, the basic state of urusovite is a crystal on valence bonds, schematically depicted in Fig. 1.9.

1.1.3. Copper trifluoroacetate

The crystal structure of the $Cu(CF_3COO)_2$ compound consists of magnetic monolayers bound by a weak van der Waals interaction, as shown in Fig. 1.10 [413]. Copper ions are in a distorted tetrahedral environment, Cu–O distances are 1.909–2.441 Å. The edge-shared tetrahedra form dimers that are connected through O–C–O bridges and form zigzag stairs. The stairs are separated by isolated CF_3 groups.

The temperature dependences of the magnetic susceptibility $\chi = M/B$ are shown in Fig. 1.11 *a*. The susceptibility at alternating current (ac) was measured with an excitation field of 10 G at a frequency of 10^4 Hz, the susceptibility at direct current (dc) was measured in the field $B = 1$ T. When the temperature is lowered, both signals pass through a wide maximum and again increase at the lowest temperatures. This behaviour is characteristic of a spin-gap system with defects/impurities. After correcting the diamagnetic contribution, the temperature dependence of χ_{ac} at high temperatures

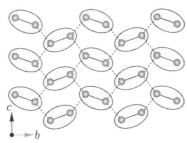

Fig. 1.9. Representation of the urusovite magnetic subsystem in the form of weakly coupled dimers against the background of a distorted lattice of the honeycomb-type shown in dashed lines [6].

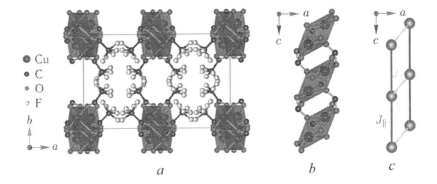

Fig. 1.10. *a*) Crystal structure of copper trifluoroacetate $Cu(CF_3COO)_2$ in a partially polyhedral representation. Elementary cell is shown by a solid line. All fluoride positions are half-filled. *b*) Dimers of Cu_2O_8. *c*) Spin-ladder of copper ions.

can be approximated by the Curie–Weiss law with a Curie constant $C = 0.43 \pm 0.05$ emu and a negative Weiss temperature $\Theta = -140 \pm 10$ K. Similar values of C and Θ within the error are obtained by analyzing the temperature dependence of χ_{dc}. Both sets of experimental data can be approximated by a model of isolated dimers with the exchange value $J \sim 200$ K. Further analysis, however, shows that the exchange interaction along the legs of the spin ladder is nonzero. An estimate of the concentration of defects/impurities was obtained from measurements of the field dependence of the magnetization M at low temperatures. As shown in Fig. 1.11 *b*, the magnetization of $Cu(CF_3COO)_2$ in strong magnetic fields is deflected downward due to the diamagnetic contribution. After correcting the diamagnetic response, the saturation magnetization makes it possible to estimate the concentration of defects/impurities as $n \sim 0.45\%$. Subtraction of the contribution of defects/impurities, shown by a dashed line, from the total signal leads to the contribution of the matrix, shown by the solid line in Fig. 1.11 *b*. Figure 1.11 *c* shows the temperature dependence of the specific heat of $Cu(CF_3COO)_2$, from which it follows that the long-range magnetic order in this compound is not reached up to $T = 2$ K.

The evolution of the electron spin resonance (ESR) spectra with temperature variation is shown in Fig. 1.12 *a*. At high temperatures, a rather intense single line of the Lorentz type is observed in the spectra, indicating an exchange narrowing. The line is characterized by the g-factor $g_1 = 2.15$, which corresponds to the signal from the ions Cu^{2+} ($S = 1/2$) in the pyramidal environment. When the temperature is lowered, the line narrows and shifts to the region of smaller magnetic fields. Two weak satellite lines with $g_2 = 4.13$ and $g_3 = 6.97$ appear when the main line disappears. The intensity of the line g_2 demonstrates the Curie–Weiss behavior at low temperatures. At the same time, the intensity of the g_3 line passes through a maximum with decreasing temperature, which is analogous to the behaviour of a system with a spin gap. At helium temperatures, a partially resolved hyperfine structure is observed in the spectra in the form of two families of four lines for parallel and perpendicular principal components of the g-tensor.

The temperature dependences of the effective g-factor and the linewidth obtained from the approximation, as well as the amplitude of the ESR signal, are shown in Fig. 1.12 *b*. The effective g-factor (upper panel) does not depend on temperature at temperatures above $T_{min} \sim 60$ K. The line width ΔB (middle panel) increases linearly with increasing temperature at $T > T_{min}$, with a slope $d(\Delta B)/dT$ of 0.1 mT/K and a residual width of the order of 28 mT. With a further decrease in temperature, ΔB passes through a minimum at T_{min} and rapidly increases at lower temperatures, whereas in the same temperature range, the value of g_{eff} increases noticeably (the top panel). The integral intensity of the ESR χ_{ESR}, which is proportional to the number of spins, was estimated by double integration of the first derivatives of the absorption curves at each temperature. In general, the ESR susceptibility is in good agreement with the data of both static and dynamic magnetic susceptibility.

The density of states (DOS) obtained from the band calculations by the generalized gradient approximation (GGA + U) method is shown in Fig. 1.13 *a*. $Cu(CF_3COO)_2$ is a dielectric with a band gap of 3 eV, the spin magnetic moment of copper Cu electrons is $0.8\mu_B$. Reduction of the magnetic moment of copper is due to the effects of

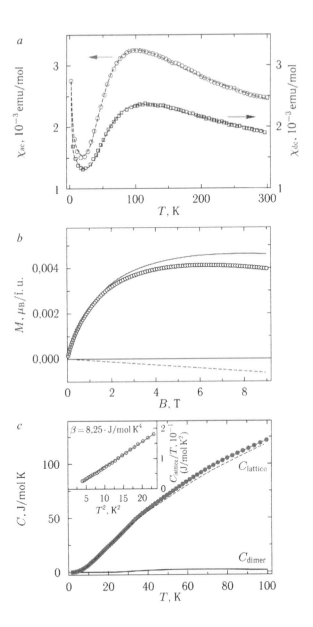

Fig. 1.11. *a*) Temperature dependences of ac- and dc-magnetic susceptibilities of $Cu(CF_3CO_2)_2$. The dotted line shows the approximation by the model of isolated dimers taking into account the temperature-independent and defects/impurities contributions. *b*) The field dependence of the magnetization of $Cu(CF_3CO_2)_2$ at $T = 2$ K. The dotted line shows the diamagnetic response. The solid curve shows the contribution of the matrix. *c*) Temperature dependence of the specific heat of $Cu(CF_3COO)_2$. The inset shows the lattice contribution at low temperatures.

hybridization with oxygen, which are characteristic of various oxides of transition metals. The upper part of the valence band is formed by O-2p states. The planar copper environment leads to the splitting of the 3d-shell, the orbital x^2-y^2 is much higher in energy than $3z^2-r^2$, and is half-filled. The corresponding peak in the single-particle density of states is observed at ~3 eV above the Fermi energy.

To calculate the exchange integrals J, the total energies of the three magnetic configurations were calculated, from which two antiferromagnetic exchange parameters $J_\perp = 176$ K and $J_\parallel = 12$ K corresponding to Cu–Cu bond lengths of 3.0974 and 3.5280 Å can be found. The larger exchange interaction $J_\perp$ occurs between the nearest copper ions forming the dimers, and $J_\parallel$ determines the interaction along the ladder guides. $J_\perp$ determines the value of the spin gap: $\Delta = 144$ K. Fig. 1.13 b depicts the distribution of the charge density corresponding to the half-filled orbital x^2-y^2.

Using a generalized quantum Monte Carlo (QMC) algorithm with exchange integrals calculated by the GGA + U method, a homogeneous magnetic susceptibility was calculated in the

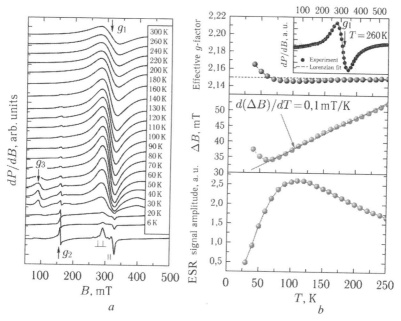

Fig. 1.12. a) Evolution of the ESR spectra with a temperature variation in the Cu(CF$_3$COO)$_2$ compound. b) Temperature dependences of the effective g-factor (upper panel), linewidth (middle panel), ESR signal amplitude (bottom panel). The inset shows an example of the ESR spectrum at high temperatures.

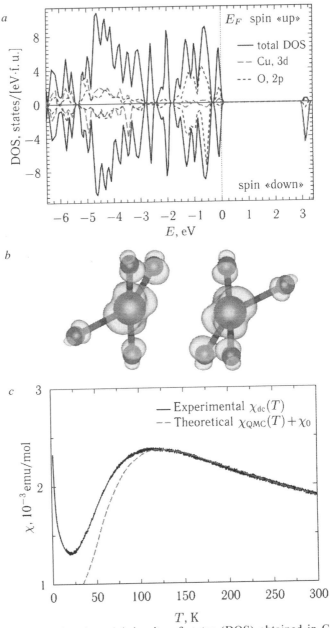

Fig. 1.13. *a*) The total and partial density of states (DOS) obtained in GGA + U calculations. Positive (negative) DOS values correspond to the direction of the spin up (down). The energy is measured from the Fermi level. *b*) The charge density corresponding to a half-filled zone, which is predominantly of x^2-y^2 origin. Two copper ions forming a dimer are shown. *c*) The magnetic susceptibility obtained in the QMC calculations of the quantum Heisenberg model with $J_\perp = 176$ K and $J_\parallel = 12$ K.

Heisenberg quantum model. As shown in Fig. 1.13 *c*, the theoretically predicted $\chi(T)$ dependence reproduces the experimental *dc*-susceptibility, the difference in the low-temperature region is due to impurities, as discussed above. At present, studies of the magnetism of organic compounds are very relevant. In particular, the question of transferring part of the magnetic moment to carbon ions is discussed. It should be noted, however, that no transfer of the magnetic moment to carbon ions connecting the copper dimers was observed [413].

1.2. Trimers

As objects where the magnetic trimers are realized on the half-integer spin $S = 1/2$, copper compounds Cu^{2+} or vanadium V^{4+} are formed as a rule. To ensure the isolation of such clusters, it is necessary to distinguish them significantly in the crystal, which implies the presence of rather complex ligands. For an equilateral triangular cluster composed of Heisenberg spins, the Hamiltonian can be written as

$$\hat{H} = J(\vec{S}_1\vec{S}_2 + \vec{S}_2\vec{S}_3 + \vec{S}_3\vec{S}_1). \tag{1.7}$$

The ground state of such a system with total spin $S_{tot} = 1/2$ is doubly degenerate and separated by a gap $\Delta = 3/2J$ from an excited fourfold degenerate state with $S_{tot} = 3/2$, as shown in Fig. 1.14.

The temperature dependence of the magnetic susceptibility of a triangular cluster is described by the expression [7]

$$\chi = \frac{N_A g^2 \mu_B^2}{4k_B T}\left(1 + 5\exp\left(\frac{3J}{2k_B T}\right)\right)\bigg/\left(1 + \exp\left(\frac{3J}{2k_B T}\right)\right). \tag{1.8}$$

Figure 1.15 (left panel) shows the temperature dependence of the effective magnetic moment in the trimer $Cu_3(O_2C_{16}H_{23})_6.1.2C_6H_{12}$. In this compound, magnetically active cations of copper are bound hrough ligands $O_2C_{16}H_{23}$ = TiBP, as shown in the right panel of Fig. 1.15. The best agreement of the theory with experiment is achieved in this case at $J = 216$ K, $g = 2.07$. A weak signal decrease at the lowest temperatures is due to intertrimer antiferromagnetic interactions, whose scale is estimated as $J' = 0.3$ K.

The magnetization curves $M(B)$ of antiferromagnetic trimers exhibit remarkable features. Figure 1.16 (upper panel) shows the $M(B)$ dependences for the $Na_9[Cu_3Na_3(H_2O)_9\ (\alpha\text{-}W_9AsO_{33})_2]$ ×

$26H_2O$ [8] system. When a pulsed magnetic field is introduced, the magnetization exhibits a plateau at $1.15\mu_B$ per unit unit, then a jump to $2.6\ \mu_B$ occurs, and finally the magnetization smoothly emerges at the saturation point of $3.4\ \mu_B$ in the field $B = 13$ T. The inset to the top panel of Fig. 1.16 shows the dependence of the magnitude of the pulsed magnetic field on time. The section AB corresponds to the input of the field, the section BC – to the field deduction and the reverse motion of the pulse to the negative field region, and the CD section – to the shutdown of the field. The high rate of change of magnetic field $\sim 10^3$ T/s is very important, since it, in fact, allows to eliminate (or, at least, noticeably weaken) the thermal exchange between the quantum spin system and the surrounding medium.

Dependence $M(B)$ demonstrates the plateau of magnetization $ngS\mu_B$ for $n = 1, 2, 3$ for $g = 2.25$. When the magnetic field is removed, the magnetization abruptly decreases to $1.15\mu_B$ and then drops to zero. When the field is expanded to positive values, the hysteresis loops uncharacteristic for copper compounds are present

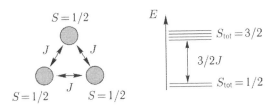

Fig. 1.14. The scheme of an isolated magnetic cluster–trimer in the form of an equilateral triangle with antiferromagnetic exchange interaction J (the left panel) and its energy diagram (right panel) [7].

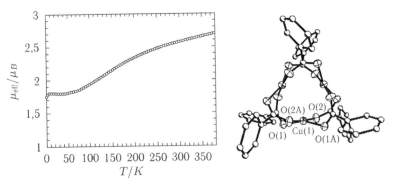

Fig. 1.15. Left panel: temperature dependence of the effective magnetic moment $Cu_3(TiBP)_6$ in the field $B = 0.1$ T. Right panel: structure of the magnetic trimer $Cu_3(TiBP)_6$. The copper and oxygen atoms are shown by ellipsoids, the carbon atoms are shown by spheres [7].

on the $M(B)$ dependence, and the magnetization is somewhat less than the equilibrium value shown by the solid line. When the field is unfolded to negative values, there are no hysteresis effects on the magnetization curve and an output to the intermediate plateau for $n = 2$. Processing of the electron paramagnetic resonance spectra in this compound made it possible to determine the value of the integral of the exchange magnetic interaction as ~4.5 K and estimate the scale of the Dzyaloshinskii–Moriya interaction as ~0.5 K. The resulting energy level diagram is shown in the lower panel of Fig. 1.16. It is shown that in the zero field there is a gap $\Delta = 1$ K between the Kramers and the second degenerate doublet, which is due to the

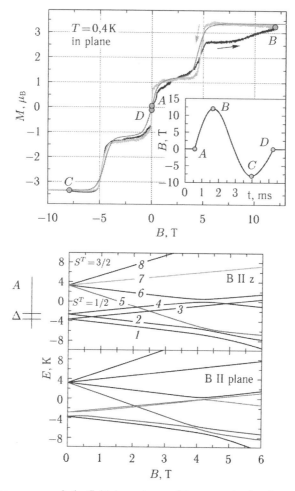

Fig. 1.16. The upper panel: the field dependence of the magnetization for $Na_9[Cu_4Na_3(H_2O)_9$ $(\alpha\text{-}W_9AsO_{33})_2] \times 26H_2O$. The lower panel: the energy level diagram in an external magnetic field, taking into account the Dzyaloshinsky–Moriya interaction [8].

magnetic anisotropy and the Dzyaloshinsky–Moriya interaction. In a zero magnetic field, the populations of levels with total spins 1/2 and −1/2 are correlated as 1: exp $\{-\Delta/k_B T\}$ = 1:0.72. Thus, the opening gap for nonequilibrium processes, for example, in the rapid sweep of the field, leads to a different population of these levels, and, consequently, to a lower value of the magnetization. In magnetic fields B = 5–8 T, intersection of the levels 1/2 and 3/2 occurs. However, only the upper level 1/2 contributes to this process, which leads to the appearance of a magnetization plateau of $2.25\mu_B$. When the magnetic field is swept into the region of negative values, the levels 1/2 are more advantageous, and the value of the moment on the plateau is $1.45\mu_B$. Thus, the ground state of an ideal triangle is characterized by a common spin S_{tot} = 1/2 with degenerate chirality (pseudospin 1/2). The Dzyaloshinsky–Moriya interaction removes this degeneracy and leads to a different behavior in magnetic field sweeps in the region of positive and negative values [8]. The possibility of using antiferromagnetic trimers as qubits controlled by the electric field of a scanning tunneling microscope was theoretically demonstrated in [9, 10].

1.2.1. Trimers of octahedra in rubidium–copper diphosphate

The monoclinic crystal structure of $Rb_2Cu_3(P_2O_7)_2$, the space group $P2_1/c$, includes linear trimers from edge-shared distorted octahedra CuO_6. As shown in Fig. 1.17, these clusters are interconnected by diphosphate groups P_2O_7 into a three-dimensional structure with channels along the c axis in which the Rb atoms are situated.

The temperature dependence of the magnetic susceptibility χ in $Rb_2Cu_3(P_2O_7)_2$, measured with cooling in a field B = 1 T, is shown in Fig. 1.18. It obeys the Curie–Weiss law at high temperatures, exhibits bending at T_N = 9.2 K, and continues to grow at lower temperatures. As shown in the left-hand inset to Fig. 1.18, below the temperature of formation of the long-range magnetic order of the $\chi(T)$ dependence, measured at field cooling (FC) and at zero field cooling (ZFC), diverge somewhat, which indicates the presence of a moderate spin disorder associated, for example, with impurities or defects. The behavior of the FC and ZFC dependences of the magnetic susceptibility at $T < T_N$ indicates the presence of a weak ferromagnetic component in the magnetic response of $Rb_2Cu_3(P_2O_7)_2$. The treatment of the experimental dependence in the high-temperature region by the Curie–Weiss law: $\chi = \chi_0 + C/$

$(T - \Theta)$, makes it possible to estimate the temperature-independent contribution $\chi_0 = -0.001$ emu/mol, the Curie constant $C = 1.185$ emu K/mol, and the Weiss temperature $\Theta = -7.5$ K. The temperature-independent contribution of χ_0 turned out to be somewhat larger than the sum of the Pascal constants. The Curie constant C corresponds to the experimental value of the square of the effective magnetic moment $\mu_{\mathrm{eff}}^2 = 9.48\mu_B^2$ per formula unit. The theoretical value of this quantity is $\mu_{\mathrm{eff}}^2 = ng^2S(S+1)\mu_B^2 = 11.29\ \mu_B^2$ for $n = 3$ numbers of magnetoactive centres Cu^{2+} ($S = 1/2$, $g = 2.24$) by the formula unit. The discrepancy between the experimental and theoretical values of μ_{eff}^2 is usually observed in low-dimensional copper oxides and is attributed to the quantum reduction of the effective magnetic moment. The negative sign of the Weiss temperature indicates the predominance of antiferromagnetic exchange interactions in the system. This is also confirmed by the temperature dependence of the reduced magnetic susceptibility $(\chi-\chi_0)$ $(T-\Theta,)$ shown in the right-hand inset to Fig. 1.18. At high temperatures this product is a horizontal line, obeying the Curie–Weiss law. Below ~ 120 K, a deviation to lower values is observed, which indicate s the predominance of antiferromagnetic correlations of short-range order. The phase transition to a magnetically ordered state at $T_N = 9.2$ K in $Rb_2Cu_3(P_2O_7)_2$ is clearly visible in this representation in the form of a maximum.

To qualitatively evaluate the scale of the antiferromagnetic correlations prevailing in the system, the $\chi(T)$ dependence in the

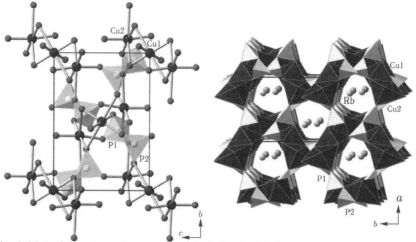

Fig. 1.17. Left panel: crystal structure of $Rb_2Cu_3(P_2O_7)_2$ in a mixed frame-polyhedral representation. Right panel: polyhedral structure view.

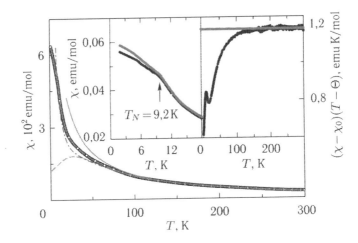

Fig. 1.18. The magnetic susceptibility of $Rb_2Cu_3(P_2O_7)_2$ in the field $B = 1$ T. The solid line represents the Curie–Weiss law (approximation from the high-temperature region). Bar-dashed lines and dotted lines correspond to different variants of the description of the magnetic subsystem. The left-hand inset shows the $\chi(T)$ dependences measured in the FC (upper curve) and ZFC (lower curve) regimes. The arrow indicates the temperature of magnetic ordering. The right-hand inset shows the temperature dependence of the reduced magnetic susceptibility $(\chi - \chi_0)$ $(T - \Theta)$.

150–300 K range was processed by the formula for the linear $S = 1/2$ antiferromagnetic trimer from [11], shown by the dash-dot line in Fig. 1.18, and the formula for the linear antiferromagnetic chain taken from [12], as shown by the dashed line in Fig. 1.18. The value of the antiferromagnetic exchange parameter in the model of isolated trimers was ~ 60 K, in the chain model ~ 40 K.

The temperature dependence of the specific heat C_p for $Rb_2Cu_3(P_2O_7)_2$, shown in Fig. 1.19, demonstrates an anomaly of the λ-type at $T_N = 9.2$ K. This anomaly corresponds to the formation of long-range magnetic order. The low-dimensional character of the magnetic subsystem manifests itself in a non-monotonic shift of the λ-anomaly when an external magnetic field is applied, as shown in the upper inset to Fig. 1.19. The total entropy S_{tot}, obtained by integrating the temperature dependence of C_p/T, shown in the lower inset to Fig. 1.19, demonstrates a kink at T_N. Below T_N, the total entropy is approximately equal to the magnetic entropy due to the small contribution of the lattice at low temperatures. The value of the entropy $\Delta S = 5.4$ J/mol K, released below T_N, can be compared with the theoretical value calculated in accordance with the formula:

$\Delta S_{\mathrm{magn}} = nR \ln (2S + 1) = 17.3$ J/mol K. Thus, the magnetic entropy released below T_N is $(\Delta S/\Delta S_{\mathrm{magn}}) \times 100\%$ ~30%. This is also characteristic of low-dimensional systems, where the magnetic entropy is released mainly at $T > T_N$ due to magnetic correlations that develop long before the transition temperature. At low temperatures, the specific heat can be described by the sum of terms proportional to T^3 and $T^{3/2}$. The dependence obtained is shown by the solid line in Fig. 1.19. The term proportional to T^3 can be related to the contribution of the phonon subsystem, the term ~$T^{3/2}$ corresponds to the contribution of three-dimensional ferromagnetic magnons.

Figure 1.20 shows the energy dependence of the density of states without spin polarization, divided by the contributions of Cu1-d, Cu2-d, O-p and P-p. It can be seen that the states at the Fermi level belong mainly to $\mathrm{Cu}-d_{x^2-y^2}$. To these states the O-p and P-p states are mixed quite strongly, which indicates the participation of P atoms in superexchange Cu–Cu interactions.

The strongest in this system were the exchange interactions between the $\mathrm{Cu}-d_{x^2-y^2}$ states involving P atoms, which turned out to be an order of magnitude greater than the interactions inside the trimer from the edge-connected Cu2–Cu1–Cu2 octahedra. Dominant the integrals of the exchange interaction J_7 together with alternating

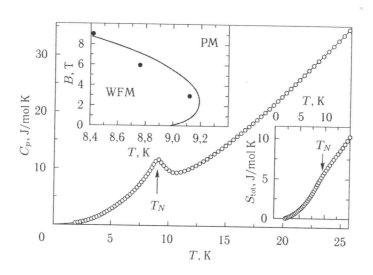

Fig. 1.19. The temperature dependence of the specific heat of $Rb_2Cu_3(P_2O_7)_2$. Upper insert: magnetic phase diagram from measurements of specific heat in an external magnetic field. The WFM region is a weakly ferromagnetic state, and PM is the paramagnetic state. Lower insert: temperature dependence of magnetic entropy,

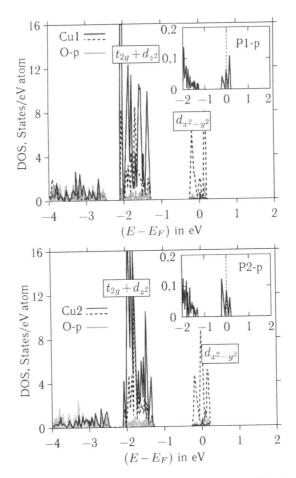

Fig. 1.20. The density of states without spin polarization, divided by the contributions of Cu1-d, Cu2-d, O-p and P-p. The energy is measured from the Fermi level E_F. The upper and lower panels show the dependences for different copper positions Cu1 and Cu2, respectively. On the inserts, contributions from the atoms P.

exchanges J_3 and J_5 form chains – Cu1–Cu2–Cu2–Cu1–, as shown on the left and right panels of Fig. 1.21. The values of the corresponding antiferromagnetic exchanges obtained using the expression $J \sim 4t^2/U$ were $J_2 = 30$ K, $J_3 = 37$ K, $J_5 = J_7 = 5$ K, at $U = 8$ eV. Thus, the spin subsystem $Rb_2Cu_3(P_2O_7)_2$ consists of antiferromagnetic spin chains connected by an additional antiferromagnetic interaction.

The low-dimensional magnetism in $Rb_2Cu_3(P_2O_7)_2$ can be understood if we take into account the details of the local environment of magnetically active Cu^{2+} ions ($S = 1/2$) and assume the predominance of antiferromagnetic exchange interactions between these ions through phosphate groups. The appreciable difference between the distances from copper to the apical oxygen atoms in the basal plane indicates that the magnetoactive orbitals $d_{x^2-y^2}$ are oriented to oxygen atoms in the basal plane. The parallel orientation of the orbitals $d_{x^2-y^2}$ leads to a significant weakening of the interactions inside the trimer, which was observed earlier in other phosphate compounds. At the same time, the treatment of magnetic susceptibility at high temperatures indicates the presence of large antiferromagnetic interactions of ~60 K. In this case, the exchange interactions between copper ions through remote apical oxygen atoms can be neglected. Then trimers of CuO_6 octahedra can not be considered as the main magnetic units in the structure of $Rb_2Cu_3(P_2O_7)_2$. The main fragments of the magnetic structure of $Rb_2Cu_3(P_2O_7)_2$ are either homogeneous chains of Cu_2 ions or alternating chains of Cu_1–Cu_2–Cu_2–Cu_1 ions interconnected through phosphate PO_4 groups, as shown in Fig. 1.21. In the latter case, the chains can be described by means of two exchange magnetic interactions in the framework of the J–J–J' model. Such chains were detected earlier in other phosphate compounds, for example, in $Cu_3(P_2O_6OH)_2$. The asymmetric bond between Cu1 and Cu2 ions with the help of a phosphate group opens the possibility for the Dzyaloshinsky-Moriya exchange interaction, which can lead to a slope of the magnetic moments, and as a result to the weak ferromagnetism observed in this compound. Another reason for

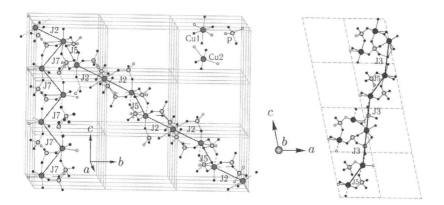

Fig. 1.21. Paths of predominant exchange interactions between Cu1 and Cu2 positions through O–P–O bridges in the structure of $Rb_2Cu_3(P_2O_7)_2$. Left panel: a homogeneous chain of Cu_2–Cu_2 formed by the exchange of J_7 and a chain of Cu1–Cu2–Cu2–Cu1 with alternating exchanges of J_2 and J_5. Right panel: the Cu1–Cu2 chain formed by the exchanges J_3 and J_5.

the weak ferromagnetic moment can be associated with frustrating interchain magnetic interactions, which form a triangular topology.

The long-range magnetic order in $Rb_2Cu_3(P_2O_7)_2$ is formed below $T_N = 9.2$ K, as follows from measurements of the magnetization and specific heat. First-principle calculations of magnetic exchanges show that exchange interactions involving P atoms predominate here. Thus, the magnetic subsystem $Rb_2Cu_3(P_2O_7)_2$ consists of interconnected antiferromagnetic spin chains, the weak ferromagnetic component in which arises from the presence of frustration and/or antisymmetric exchange interactions.

1.2.2. Trimer plaquettes in sodium–copper germanate

The compound with the triclinic symmetry of $Na_3Cu_3Ge_4O_{12}$ crystallizes in the space group $P\bar{1}$ [13]. In the $Na_3Cu_3Ge_4O_{12}$ structure, Na^+ ions are located in tunnels formed by six polyhedra (two CuO_4 squares and four GeO_4 tetrahedra) connected vertically. As shown in the right panel of Fig. 1.22, edge-shared squares of CuO_4 form a linear isolated trimer of Cu_3O_8. The trimers are located one above the other in the [001] direction. The Cu–O bond length inside the dimer is 1.92–1.98 Å, the Cu(1)–O–Cu(2) bonding angles in the trimer are equal to 100.0–102.2°, the Cu–Cu distance is small and equal 3.03 Å. The bond between the trimers is much weaker, the Cu–O distance is 2.694 Å, the distances between copper ions in neighbouring trimers shown in dotted lines in the right panel of Fig. 1.22, are 3.349 Å for Cu(1)–Cu(2) and 3.273 Å for Cu(2)–Cu(2) [13]. The angles of the Cu–O–Cu bonds between the trimers are 91.4°–94.6°. The corresponding picture of the magnetic interactions is shown in Fig. 1.23 [14], where S_1, S_2, S_3 are the spins of the ions forming the trimer, J_1 is the exchange between the nearest neighbors, J_2 is the exchange between neighbors next to the nearest ones, and J_3 is the exchange between neighboring trimers.

Figure 1.24 shows the temperature dependence of the magnetic susceptibility of $Na_3Cu_3Ge_4O_{12}$ in the ZFC regime, $B = 0.1$ T [14]. A broad maximum is observed at $T \sim 11$ K, associated with the formation of a regime of short-range correlations. With further cooling, a sharp decrease in the magnetic susceptibility is observed at $T_C = 2$ K, as shown in the inset to Fig. 1.24. This decline is associated with the formation of a long-range magnetic order, which is confirmed by the anomaly in the specific heat of $Na_3Cu_3Ge_4O_{12}$, shown in Fig. 1.25 [14].

To analyze the behaviour of the magnetic susceptibility in the high-temperature region 70–650 K, a description was used for a trimer using the Heisenberg model. In the case of the spin projections $S_1^z = S_2^z = S_3^z = \pm 1/2$, this problem can be reduced to the problem of diagonalization and calculation of the eigenvalues and eigenfunctions of the matrix of 8×8 elements, each of which is a base vector in the form $|\uparrow\uparrow\uparrow\rangle$, $|\uparrow\uparrow\downarrow\rangle$, etc. As a result, the magnetic susceptibility can be described by the equation

$$\chi(T) = \frac{N_A g^2 \mu_B^2}{12 k_B T} \times$$

$$\times \frac{10\exp(-J_1/2k_B T) + \exp(J_1/k_B T) + \exp(J_2/k_B T)}{2\exp(-J_1/2k_B T) + \exp(J_1/k_B T) + \exp(J_2/k_B T)} + \chi_0. \tag{1.9}$$

The result of processing the magnetic susceptibility according to this formula is shown in Fig. 1.24 with a dotted line. According to this estimate, the values of exchange integrals in a trimer are:

$$J_1 = 30 \pm 20 \text{ K}, \quad J_1 = 340 \pm 20 \text{ K}$$

(both antiferromagnetic), the temperature-independent term in the magnetic susceptibility is $\chi_0 = 6 \cdot 10^{-5}$ emu/mol Cu. The ground state of the trimer is a doublet of wave functions:

$$\phi_0^A = \frac{1}{\sqrt{2}}\left(|\uparrow\downarrow\downarrow\rangle - |\downarrow\downarrow\uparrow\rangle\right), \quad \phi_0^B = \frac{1}{\sqrt{2}}\left(-|\uparrow\uparrow\downarrow\rangle + |\downarrow\uparrow\uparrow\rangle\right). \tag{1.10}$$

It can be seen that in the ground state the spins at the edges of the trimer (designated S_1 and S_3 in Fig. 1.23) are connected by a strong antiferromagnetic exchange J_2 and form a spin singlet. The central spin S_2 remains free, since the action of the two exchanges with S_1 and S_3 is compensating each other. The gap between the ground state and the excited state can be estimated as $\Delta = 310$ K $(= J_2 - J_1)$, i.e., the ground state is stable at low temperatures. To describe the behaviour of the magnetic susceptibility, it is necessary to consider the exchange integral J_3, which connects the spins S_2 in neighbouring trimers. As shown in Fig. 1.23, the spins S_2 form one-dimensional chains, the susceptibility behaviour for which can be described by the Bonner–Fisher model. The res0ult of modelling the magnetic susceptibility in the interval $T_C < T < 70$ K is shown in Fig. 1.24 by

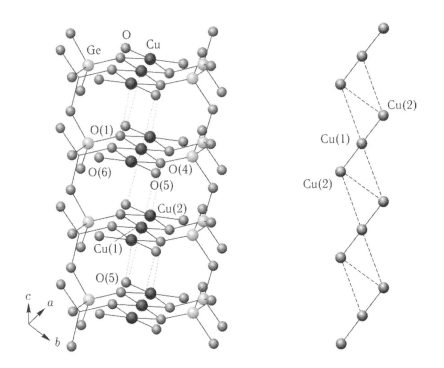

Fig. 1.22. The crystal structure and copper–oxygen bonds in the compound $Na_3Cu_3Ge_4O_{12}$ [13].

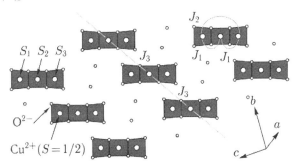

Fig. 1.23. Schematic representation of the exchange interaction paths in $Na_3Cu_3Ge_4O_{12}$ [14].

a solid line, the J_3 estimate for this model was $J_3 = 18$ K. However, the one-dimensional chain model can not describe the phase transition at T_C, since the long-range order is not formed in the uniform chain $S = 1/2$ even at the lowest temperatures. The order at $T_C = 2$ K can arise only due to the interaction between the chains [13].

Analysis of the specific heat in the range 4–20 K was also carried out according to the Bonner–Fisher model [12]. The specific heat

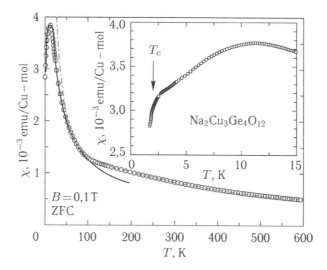

Fig. 1.24. The temperature dependence of the magnetic susceptibility of the compound Na$_3$Cu$_3$Ge$_4$O$_{12}$ [14].

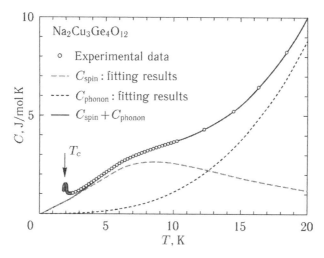

Fig. 1.25. The temperature dependence of the specific heat of Na$_3$Cu$_3$Ge$_4$O$_{12}$ [14].

was divided into two contributions. The phonon contribution is $\sim\beta T^3$, where a value of $\beta = 1.1 \cdot 10^{-3}$ is used to estimate the Debye temperature of 330 K. The magnon contribution was determined on the assumption that the exchange parameter along the chain is $J_3 = 18$ K chain, as follows from the temperature dependence of the magnetic susceptibility.

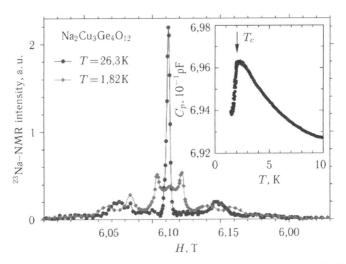

Fig. 1.26. The intensity of the signal of nuclear magnetic resonance in $Na_3Cu_3Ge_4O_{12}$ at temperatures of 26.3 and 1.82 K. On the inset: the temperature dependence of the dielectric constant [14].

In [14], nuclear magnetic resonance spectra were obtained on ^{23}Na nuclei at temperatures of 26.3 and 1.82 K, as shown in Fig. 1.26. The shape of the spectrum changes dramatically below T_c, which is due to a change in the local field on Na, which occurs when the long-range magnetic order is formed in the copper subsystem. The NMR spectrum in $Na_3Cu_3Ge_4O_{12}$ at 1.82 K resembles the NMR spectrum in $NaCu_2O_2$ [15], in which the existence of a spiral magnetic structure that is not commensurate with the lattice is proved (see Section 2.2.2). The inset to Fig. 1.26 shows the temperature dependence of the condenser capacitance, where $Na_3Cu_3Ge_4O_{12}$ is used as a dielectric. The value of the capacitance C_p is directly proportional to the dielectric constant of $Na_3Cu_3Ge_4O_{12}$. The transition at $T_c = 2$ K is accompanied by an anomaly in the dielectric constant in the form of a peak, but the authors of [14] did not investigate the field dependence of the polarization $P(E)$ at low temperatures, which would be a direct proof of the multiferroelectric behaviour of $Na_3Cu_3Ge_4O_{12}$. Nevertheless, we can conclude that the trimer system $Na_3Cu_3Ge_4O_{12}$ is a multiferroic below 2 K due to the magnetic structure, which is not commensurate with the lattice.

1.3. Tetramers

Among the family of isolated tetramers with spin $S = 1/2$, systems where the magnetic centres form plaquettes of CaV_4O_9 [16] and

$NaCuAsO_4$ [17], as well as extended linear structures, for example, in $SrCu_2(PO_4)_2$ [18] have been identified. In all these compounds a spin gap is present in the spectrum of magnetic excitations. A distinctive feature of the latter phosphate is a magnetic tetramer formed due to the interaction between the magnetic ions Cu^{2+} through the intermediate phosphate groups PO_4. It should be noted that the magnetic exchange interactions through the intermediate groups PO_4 or AsO_4 often play a more important role than the magnetic exchange along the metal–oxygen–metal path.

Of great interest, both for the experiment and for model calculations is the CaV_4O_9 compound. This compound has a tetragonal crystal lattice, the space group *P4/n*. In this structure, the vanadium magnetic ions V^{4+} ($S = 1/2$) surrounded by five oxygen ions are disposed in layers, and within the layer form a system of four-ion clusters – plaquettes interacting with each other.

The Hamiltonian describing the interactions in such a system must account for two types of magnetic exchange inside the plaquette for the *i*th ion: 2 nearest neighbours on the plaquette (J_1) and 1 neighbour diagonally (J_2), and two exchanges between neighbouring plaquettes: 1 nearest ion in the next plaquette (J_1') and 2 neighbours diagonally from the other plaquettes (J_2'):

$$H = J_1 \sum_{(i,j)} S_i \cdot S_j + J_1' \sum_{(i,k)} S_i \cdot S_k + J_2 \sum_{(i,l)} S_i \cdot S_l + J_2' \sum_{(i,m)} S_i \cdot S_m. \tag{1.11}$$

The vanadium ion V^{4+} has only one electron per group of five *d*-orbitals, so CaV_4O_9 is described using a square lattice model with bonds filled by 1/5. Figure 1.27 shows the model scheme for the system of plaques.

From the point of view of modelling of magnetic properties, this system proved to be very attractive for numerical simulations, the number of published articles with calculations is ten times higher than the number of experimental works. Depending on the magnitudes and signs of the exchange integrals shown in Fig. 1.27, several regions of the phase diagram for plaquette systems can be distinguished, in particular, there is a region corresponding to the dimerized state, where dimers can be formed by ions from one plaque or by ions of adjacent plaquettes. There are also several variants of the formation of a sloping antiferromagnetic order, a state of the spin–Peierls type or a disordered state can be realized. In a real substance, it was found that when the temperature is lowered, the magnetic susceptibility

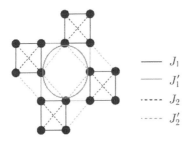

Fig. 1.27. Scheme of exchange integrals for a two-dimensional system of interacting plaquettes. The meta-plaquette is shown by an oval [19].

passes through a wide maximum and vanishes in all directions, as shown in Fig. 1.28 [16]. This behaviour corresponds to a non-magnetic ground state with a spin gap $\Delta \sim 107$ K.

Initially, it was assumed that in the ground state the system can be represented as simple plaquettes that are interconnected. However, the peculiarities in the arrangement of vanadium ions in the and the path of the exchange interaction do not agree with the simple picture, in particular, because of the asymmetric arrangement of the ions of the oxygen environment, the orbital d_{xy} turns out to be the lowest in energy, and it is the only electron V^{4+} that is located on it. The values of the exchange integrals for the real object are $J_1 = J_1' = 78.9$ K, $J_2 = 19.7$ K, $J_2' = 162.5$ K. Thus, the main interaction is the exchange J_2' and the spin-singlet state is formed on a meta-plaquette of 4 ions belonging to different plaquettes, as shown in Fig. 1.27 [19].

The thermodynamic properties of tetramers $S = 1/2$ can be considered using the example of $SrCu_2(PO_4)_2$. This compound has a triclinic crystal structure, the space group $P\bar{1}$. The structure of the magnetic tetramer and the temperature dependence of the magnetic susceptibility are shown in Fig. 1.29. Copper ions Cu^{2+} are in the pyramids $Cu1O_5$ and squares $Cu2O_4$, which are connected through one or two phosphate groups.

A wide maximum on the $\chi(T)$ dependence is replaced by a fall in the magnetic susceptibility. practically to zero. The growth of the magnetic susceptibility at low temperatures is due to the presence of impurity centres. The linear alternating tetramer Hamiltonian can be written as:

$$\hat{H} = -J_1(\vec{S}_1 \cdot \vec{S}_2 + \vec{S}_3 \cdot \vec{S}_4) - J_2(\vec{S}_2 \cdot \vec{S}_3). \qquad (1.12)$$

The solution of such a Hamiltonian is a system of six energy levels:

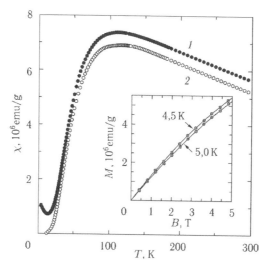

Fig. 1.28. Magnetic susceptibility of CaV$_4$O$_9$. Curve 1 – experiment, curve 2 – after subtraction of the paramagnetic contribution of defects. In the inset: the field dependence of the magnetization at two temperatures [16].

$$E_1(S=2) = -J_1/2 - J_2/4,$$
$$E_2(S=1) = J_1/2 - J_2/4,$$
$$E_3(S=1) = J_2/4 + 1/2\sqrt{J_1^2 + J_2^2},$$
$$E_4(S=1) = J_2/4 - 1/2\sqrt{J_1^2 + J_2^2},$$
$$E_5(S=0) = J_1/2 + J_2/4 + \sqrt{J_1^2 - 1/2 J_1 J_2 + 1/4 J_2^2},$$
$$E_6(S=0) = J_1/2 + J_2/4 - \sqrt{J_1^2 - 1/2 J_1 J_2 + 1/4 J_2^2}.$$

(1.13)

In this case, the temperature dependence of the magnetic susceptibility of the tetramer is described by the equation [16]

$$\chi = \frac{Ng^2\mu_B^2}{4k_BT}\left(10\exp\left(-\frac{E_1}{k_BT}\right) + 2\exp\left(-\frac{E_2}{k_BT}\right) + \right.$$

$$2\exp\left(-\frac{E_3}{k_BT}\right) + 2\exp\left(-\frac{E_4}{k_BT}\right)\Bigg/\left(5\exp\left(-\frac{E_1}{k_BT}\right) + 3\exp\left(-\frac{E_2}{k_BT}\right) + \right.$$

$$3\exp\left(-\frac{E_3}{k_BT}\right) + 3\exp\left(-\frac{E_4}{k_BT}\right) + \exp\left(-\frac{E_5}{k_BT}\right) + \exp\left(-\frac{E_6}{k_BT}\right)\Bigg).$$

(1.14)

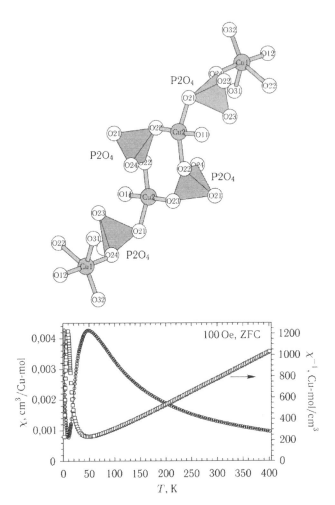

Fig. 1.29. The upper panel: the device of the magnetic linear tetramer. Bottom panel: temperature dependences of the direct and inverse magnetic susceptibility of $SrCu_2(PO_4)_2$. The line shows processing in the model of linear tetramers [18]

The dependence of $\chi(T)$ on this formula, summed with the contribution of impurity centres, is shown in Fig. 1.29 by a solid line. The exchange magnetic interactions in the tetramer amounted to $-J_1 = 82$ K and $-J_2 = 59$ K. The gap in the spectrum of magnetic excitations, defined as $\Delta = E_4 - E_6$, was 63 K. An investigation of the ^{63}Cu nuclear quadrupole resonance spectra confirmed the presence of a gap in this compound $\Delta = 65$ K [18]. The field dependence of the magnetization of $SrCu_2(PO_4)_2$, shown in Fig. 1.30, demonstrates almost linear growth at $B > B_1 = 28$ T. The spin gap can be determined from the relation $\Delta_1/k_B = 0.6714gB_1 = 42$ K for $g = 2.2$. In the interval 50–63 T, the $M(B)$ dependence shows the plateau 1/2, which is associated with intercluster interactions.

1.4. Bose–Einstein condensation of magnons

Bose–Einstein condensation is one of the most interesting phenomena predicted by quantum mechanics. This phenomenon assumes the formation of a collective quantum state consisting of identical particles with an integer angular momentum or spin (bosons), when the particle density exceeds the critical value [20].

The concept of a magnon was first introduced by Bloch to describe the reduction of the magnetization of a ferromagnet at a temperature different from zero. Each magnon reduces the total spin of the system by one $\hbar$ or decreases the magnetization by gh, where g is the gyromagnetic ratio [21]. In [22], the exact correspondence between the quantum antiferromagnet and the Bose gas system was shown. In particular, the analogy between spins and bosons takes place in antiferromagnetic dimer systems, where closely spaced pairs of spins $S = 1/2$ form a spin-singlet $S = 0$ and a triplet $S = 1$ state. Here boson excitations are called triplons, which are close to magnetic excitations of an ordered antiferromagnet or magnons. In particular, they have the same quantum numbers, so these two terms are sometimes used as synonyms.

The Bose–Einstein condensation of magnons was observed in the spin-dimeric compounds $TlCuCl_3$ and $BaCuSi_2O_6$ [23]. The lattice of magnetic ions in such materials is a set of dimers composed of Cu^{2+} ions, which are bound by intradimer and interdimer antiferromagnetic interactions J_0, J_{mnij}:

$$\hat{H} = \sum_i J_0 S_{1,i} S_{2,j} + \sum_{mnij} J_{mnij} S_{m,i} S_{n,j} - g\mu_B B \sum_{mi} S^z_{m,j}, \qquad (1.15)$$

where B is the external magnetic field in the z direction; i, j is the number of magnetic dimers, $m, n = 1, 2$ are their magnetic states. For a square lattice composed of similar dimers, the energy of triplets is defined as:

$$\varepsilon(\vec{k}) = J_0 + J_1[\cos(k_x a) + \cos(k_y a)] - g\mu_B B S^z, \qquad (1.16)$$

where $\mathbf{k} = (k_x, k_y)$ denotes the wave vector, a is the lattice constant,

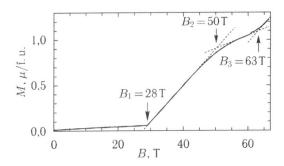

Fig. 1.30. The field dependence of the magnetization of $SrCu_2(PO_4)_2$ at 1.3 K [18].

and $D = 4J_1$ is the width of the band determined by the interdimer interaction.

The Zeeman contribution $-g\mu_B BS^z$ – determines the density of triplons. When an external magnetic field is applied in spin-dimer systems, the triplon level with $S^z = +1$ is dropped and, in the end, crosses zero, as shown in the upper panel of Fig. 1.31. This determines the two critical magnetic fields B_{C1} and B_{C2} in the phase diagram, as shown in the bottom panel of Fig. 1.31. At zero temperature in fields below B_{C1}, the magnetization is zero and only singlets exist. Between B_{C1} and B_{C2} the magnetization increases with increasing field, as more triplons appear in the ground state due to the increased gain in the Zeeman energy. Above the field B_{C2} there are only triplons, and the magnetization goes to saturation. Near B_{C1} and B_{C2}, the phase boundary must follow the power law $T_C \sim (B-B_{C1})^\phi$ with a universal power exponent $\phi = 2/3$ in accordance with theory [23].

1.4.1. Pigment of the Han Dynasty

Experimental studies of the Bose–Einstein condensation of magnons in spin–dimer systems indicate deviations from the power-law field dependence of the critical temperature. So, in $BaCuSi_2O_6$, as shown in Fig. 1.32, copper ions in a square oxygen environment CuO_4, connected through silicate groups, form dimers $S = 1/2$ [24]. This compound has a tetragonal crystal structure, space group $I4_1/acd$. From the magnetic characterization data and absorption spectra of neutron radiation, the value of the magnetic gap is established as 4.5 meV (52 K). For this compound, the boundaries of existence of phases on the magnetic phase diagram are established. However, at a temperature $T = 0.5$ K, there is a change in the power dependence

from $\phi = 2/3$ at $T > 0.5$ K to $\phi = 1$ at $T < 0.5$ K. It was shown in
[25] that these exponents are characteristic for three-dimensional and
two-dimensional Bose–Einstein condensates. This reduction in the
dimension of the Bose condensate was associated with the frustration
of interplane interactions.

1.4.2. Barium–vanadium disilicate

The tetragonal crystal structure of $BaVSi_2O_7$ (the space group $I4/m$

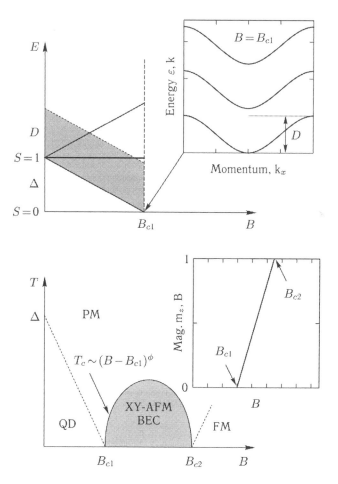

Fig. 1.31. Top panel: Zeeman splitting of triplet modes separated by a gap Δ,
with energy D for $k_0 = (\pi/a, \pi/a)$. The inset shows the dispersion of triplons at the
critical field B_{c1}. Bottom panel: the resulting phase diagram with paramagnetic
(PM), field-induced ferromagnetic (FM) and the 'canted' antiferromagnetic (XY
AFM) regions [23].

[26]) is shown in Fig. 1.33 [27]. Magnetoactive cations V^{4+} are in a distorted pyramidal oxygen environment. The pyramid VO_5 is formed by four basal and one apical oxygen ions. The d_{xy} orbitals of the vanadium ion are magnetoactive. Pyramids VO_5 constitute bipyramidal dimers connected through silicate groups SiO_4, associated with each basal oxygen ion, as shown in Fig. 1.33 (right panel).

The magnetic interaction inside the J_d dimers is organized through V–O–O–V bonds. The complexes of vanadium dimers consisting of two VO_5 pyramids and four SiO_4 tetrahedra are connected through vertices of silicate groups, forming a two-layered square lattice. Neighbouring bi-layers are shifted relative to one another for half a period. The temperature dependence of the magnetic susceptibility of $BaVSi_2O_7$, measured in the field $B = 0.1$ T, is shown in Fig. 1.34. At high temperatures, the $\chi(T)$ dependence obeys the Curie–Weiss law with a temperature-independent contribution $\chi_0 = -2.0 \cdot 10^{-5}$ emu/mol, the Curie constant $C = 0.34$ emu K/mol, and the Weiss temperature $\Theta = -9.2$ K. The experimentally established Curie constant C gives an underestimate for the square of the effective magnetic moment $2.7\mu_B^2$ in comparison with the expected value $\mu_{eff}^2 = 3.0\mu_B^2$. The value of the Weiss temperature makes it possible to estimate the exchange interaction parameter between the nearest V^{4+} ions in $BaVSi_2O_7$ as $J_d = 37$ K.

With decreasing temperature, the $\chi(T)$ dependence deviates from the Curie–Weiss law, passes through a wide maximum at $T_{max} = 23$ K, and rapidly decreases. At the lowest temperatures, the $\chi(T)$ curve shows a weak rise, due to the presence of defects/impurities in the sample. Throughout the investigated temperature range the magnetic susceptibility can be described by the sum of the temperature-independent contribution χ_0, the impurity contribution of χ_{imp}, which obeys the Curie law, and the dimer contribution of χ_{dim}, which can be described in the model of non-interacting dimers.

The field dependence of the magnetization of $BaVSi_2O_7$ was measured in a pulsed magnetic field at $T = 1.6$ K, as shown in Fig. 1.35. The main feature of this dependence is a rapid increase in magnetization at $B_C = 27.2$ T, which manifests itself as a maximum on the derivative, beyond which saturation is practically reached. This feature indicates a transition to a triplet state, i.e., the destruction of intradimer interaction by an external magnetic field. The exchange interaction parameter can be determined from the value of the critical field as $J_d = g\mu_B B_C = 36$ K, which agrees with the value obtained from the temperature dependence of the magnetic susceptibility.

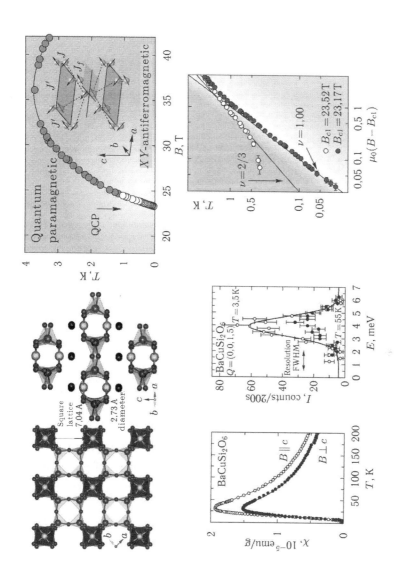

Fig. 1.32. The left panel: a fragment of the crystal lattice, the temperature dependences of the magnetic susceptibility and scans at constant Q in BaCuSi$_2$O$_6$ [24]. The right panel: the phase diagram and the field dependence of the critical temperature in BaCuSi$_2$O$_6$ [25].

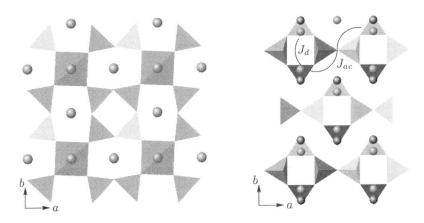

Fig. 1.33. The crystal structure of BaVSi$_2$O$_7$ [27]. Left panel: layer topology, right panel: the layout of several layers. Polyhedra are VO$_5$ pyramids and SiO$_4$ tetrahedrons. Ba^{2+} ions as separate spheres are located between layers

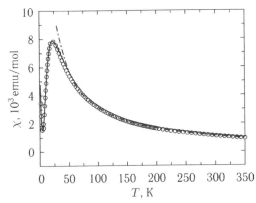

Fig. 1.34. The temperature dependence of the magnetic susceptibility of BaVSi$_2$O$_7$ [27]. Bar–dotted lines are the extrapolation of the Curie–Weiss law, the solid line is the sum of all contributions.

In weak magnetic fields, excess magnetization is observed, which indicates the contribution of impurity magnetic moments.

The dependence $M(B)$ can be approximated in the model of non-interacting dimers by the formula:

$$M = \frac{g\mu_B}{Z}\left[\exp\left(-\frac{J_d - g\mu_B B}{k_B T}\right) - \exp\left(-\frac{J_d + g\mu_B B}{k_B T}\right)\right], \qquad (1.17)$$

where Z is the partition function defined by the expression:

$$Z = 1 + \exp\left(-\frac{J_d - g\mu_B B}{k_B T}\right) + \exp\left(-\frac{J_d}{k_B T}\right) + \exp\left(-\frac{J_d + g\mu_B B}{k_B T}\right). \quad (1.18)$$

Approximation of the $M(B)$ dependence at $T = 1.6$ K with the parameters $J = 36$ K and $g = 1.975$ is shown by the solid line in Fig. 1.35. It can be seen that in the intermediate fields ($B < B_c$), the experimentally observed magnetization values slightly exceed the contribution of defects/impurities, as well as the contribution from dimers in the model of non-interacting dimers. Interdimer exchange interactions can lead to excess magnetization in BaVSi$_2$O$_7$ at $B < B_C$.

The analysis of the static and dynamic magnetic properties given above is confirmed by the temperature dependence of the specific heat in BaVSi$_2$O$_7$ shown in Fig. 1.36. Obviously, there is an appreciable additional contribution to the specific heat at low temperatures, in comparison with the specific heat in the isostructural non-magnetic analog of BaTiSi$_2$O$_7$. This additional contribution can be attributed to an anomaly of the Schottky type associated with singlet–triplet excitations in antiferromagnetic dimers.

The obtained experimental results show that the magnetic properties of BaVSi$_2$O$_7$ can be explained in the model of weakly interacting dimers. To test this scenario, calculations of the electronic structure were performed using a modified method of local density approximation (LDA) [28, 29]. The resulting band structure and

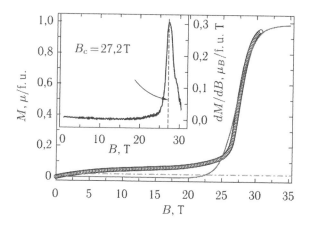

Fig. 1.35. The magnetization curve is BaVSi$_2$O$_7$ at $T = 1.6$ K. The contribution of defects/impurities is approximated by the Brillouin function shown by the dash-dot line. On the inset: the derivative of the magnetization curve; the critical magnetic field B_C is indicated by a vertical dashed line [27].

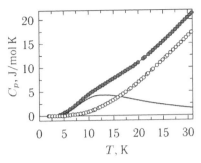

Fig. 1.36. Temperature dependences of the specific heat BaVSi$_2$O$_7$ (closed symbols) and its isostructural nonmagnetic analog BaTiSi$_2$O$_7$ (open symbols). The contribution of dimers with the exchange of J_d = 37 K is shown by a solid line [27].

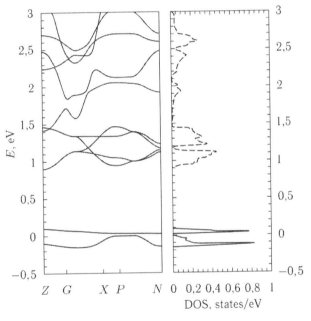

Fig. 1.37. Band structure (left panel) and partial density of states (right panel), obtained from LDA calculations [27]. Solid and dashed lines denote the density of states $3d_{xy}$ and the total density of $3d$ states, respectively. The Fermi energy is zero.

the partial density of states are shown in Fig. 1.37. Near the Fermi level, there are two well-separated zones of binding and loosening orbitals. In accordance with the calculated partial density of states presented in the right panel of Fig. 1.37, these bands correspond to $3d$ states of vanadium with symmetry xy. The lowest excited states have yz and xz symmetry. This structure of the LDA spectrum is characteristic for vanadium S = 1/2 oxides [30].

It is of interest to compare the electronic structures of $BaVSi_2O_7$ and $BaCuSi_2O_6$ compounds. A noticeable difference between them is the degree of hybridization of the transition metal and oxygen states. For the $BaCuSi_2O_6$ system, the ground (magnetic) state is associated with the e_g orbital with the symmetry x^2-y^2, which is directed to the ligand oxygen atoms. In the case of $BaVSi_2O_7$, the magnetic orbitals of the t_{2g} group are not directed to oxygen. This difference leads to strong exchange interactions of J_{ac} between the dimers in the copper compound. In addition, the magnetic moment of the copper atom decreases, which is associated with strong hybridization with oxygen [16]. The calculated value of the main antiferromagnetic exchange of the interaction inside the dimer is $J_d = 52$ K, which is somewhat higher than the experimentally determined value of 37 K. The ferromagnetic interaction between the dimers leads to an excess magnetization in $BaVSi_2O_7$ observed on the $M(B)$ dependence at $B < B_c$. Thus, the model of weakly interacting dimers in $BaVSi_2O_7$ is confirmed from the first-principles calculations.

Quasi-one dimensional magnets

2.1. Homogeneous chain of half-integer spins

The antiferromagnetic chain of the spins $S = 1/2$ is described by the Hamiltonian:

$$\hat{H} = J\sum_i S_i S_{i+1}, \tag{2.1}$$

where $J > 0$ denotes the antiferromagnetic exchange between the nearest neighbours. Such a system admits an exact solution, that is, its eigenvalues can be obtained from the Bethe equations [31]. The first attempts to describe the thermodynamic properties of such a system were made in [12, 32], where a high-temperature region was described quite well.

The calculation of the magnetic susceptibility and specific heat of the Heisenberg N-spin ring was made in [12]. The curves obtained in the limit $N \rightarrow \infty$ are shown in Fig. 2.1. The magnitude and position of the wide maximum on the curve $\chi(T)$ is determined by the relations:

$$\frac{\chi_{max}|J|}{g^2\mu_B^2} \approx 0.7346, \qquad \frac{kT_{max}}{|J|} \approx 1.282. \tag{2.2}$$

The parameters of the broad maximum on the temperature dependence of specific heat are determined by the relationships:

$$\frac{C_{max}}{Nk_B} \approx 0.35, \qquad \frac{kT_{max}}{|J|} \approx 0.962. \tag{2.3}$$

When processing the experimentally obtained values of the temperature dependence of the magnetic susceptibility, it is convenient to use analytical expressions. The $\chi(T)$ dependence for a homogeneous antiferromagnetic chain $S = 1/2$ over a wide range of temperatures can be represented by a polynomial:

$$\chi(T) = \frac{Ng^2\mu_B^2}{k_BT} \frac{A_0 + A_1\left(\dfrac{J}{2k_BT}\right) + A_2\left(\dfrac{J}{2k_BT}\right)^2}{1 + B_1\left(\dfrac{J}{2k_BT}\right) + B_2\left(\dfrac{J}{2k_BT}\right)^2 + B_3\left(\dfrac{J}{2k_BT}\right)^3}, \tag{2.4}$$

where the values of the parameters A_i and B_i are given in [33].

In general, the ground state of a homogeneous antiferromagnetic chain is a gapless spin liquid, and such a chain should not undergo magnetic ordering up to $T = 0$. The long-range order in a chain of atoms bound by the exchange interaction J is destroyed by a single-spin flip. In this case, the magnetic energy increases by $2J$, and the entropy increases by $k_B \ln N$. The change in the free energy for a spin flip is written as: $\Delta F = 2J - k_BT \ln N$ and can be made negative for any arbitrarily low temperature, by choosing a sufficiently large the number of links in the chain N [34].

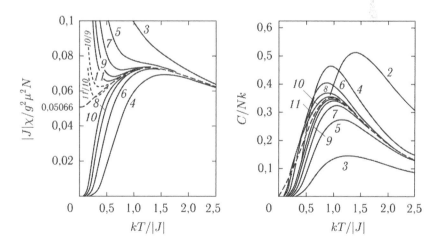

Fig. 2.1. Temperature dependences of the magnetic susceptibility (left panel) and specific heat (right panel) for Heisenberg antiferromagnetic clusters. The numbers denote the number of links in clusters, the case of an infinite chain $N \to \infty$ is represented by dashed lines [12]. The above graphs are taken from the original article, where the calculations were performed in writing the Hamiltonian via $2J$.

2.1.1. Rubidium–copper molybdate

In the vast majority of quasi-one dimensional compounds with large values of the exchange interaction constants at a certain temperature, three-dimensional magnetic ordering occurs due to the interchain interaction. One of the metal oxide compounds, where the properties of the homogeneous antiferromagnetic Heisenberg chain are realized most clearly, is $Rb_4Cu(MoO_4)_3$.

The compound $Rb_4Cu(MoO_4)_3$ at room temperature has an orthorhombic crystal structure, the space group *Pnma* [35]. In this compound, the magnetically active copper ions $Cu^{2+}(d_{x^2-y^2}, S = 1/2)$ are in the pyramidal environment of oxygen anions. Pyramids, joined together by MoO_4 groups, form two-dimensional planes *ac*, as shown in the left upper panel of Fig. 2.2. In this case, the magnetic pyramids can interact with each other through intermediate molybdenum groups along paths J and J'. The bond J is organized

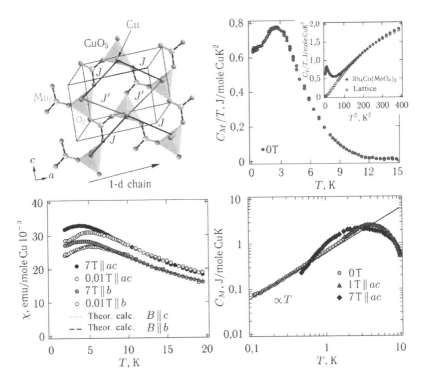

Fig. 2.2. Left panels: a fragment of the crystal structure and temperature dependences of the magnetic susceptibility of $Rb_4Cu(MoO_4)_3$. Right panels: temperature dependences of specific heat C_p and magnetic contribution of C_M for $Rb_4Cu(MoO_4)_3$ at different temperature intervals [35].

through the basal oxygen atoms in the pyramid CuO_5, then in the case of J' the molybdenum group coordinates the basal and apical oxygen atoms in the neighbouring pyramids, which significantly weakens this exchange.

The dependence of the magnetic susceptibility of $Rb_4Cu(MoO_4)_3$ on temperature has a wide maximum at low temperatures. The theoretical treatment in the model of homogeneous antiferromagnetic chains is represented by a dashed line and agrees qualitatively with the experimental data for $J = 10$ K. The experimental values of the parameters of the maximum χ_{max} and T_{max} are close to their estimates by formula (2.2). The temperature dependence of specific heat of this compound also has a wide maximum, as shown in the right panel of Fig. 2.2, which is manifested more clearly for the magnetic specific heat C_M after subtracting the lattice contribution using a non-magnetic analog of $Rb_4Zn(MoO_4)_3$. There are no indications of the transition to a magnetically ordered state on the $C_M(T)$ dependence up to 0.1 K. The experimental values of the maximum parameters in specific heat C_{max} and T_{max} are close to the model of a homogeneous antiferromagnetic chain described by the formula (2.3) [35].

2.1.2. Vanadyl diacetate

According to the crystallographic data [36], vanadyl diacetate $VO(CH_3COO)_2$ has a non-centrosymmetric $Cmc2_1$ space group. The structure contains the square pyramids VO_5, interconnected through apical oxygen (V = O 1.684 (7) Å, V...O 2.131 (7) Å). Thus, in the structure there are layers from infinite one-dimensional chains V = O...V = O..., as shown in Fig. 2.3. The pyramid bases consist of four oxygen atoms belonging to the acetate groups $[CH_3COO]^-$ with distances V–O 1.931 (4) Å and 2.002 (3) Å. As shown on the right-hand side of Fig. 2.3, alternating acetate bridges curve the chain in such a way that V = O...V angle is 131.2 (3) °. At the same time, the pyramid itself is distorted little, the angle O = V...O is equal to 174.6 (2)°. Magnetoactive chains in the vanadyl diacetate structure are far apart, suggesting a weak magnetic coupling between them.

The primary measurements of the magnetic susceptibility of $VO (CH_3COO)_2$, obtained in Ref. [36], exhibit a broad maximum for the susceptibility of 1D chains on the $\chi(T)$ dependence at $T = 275$ K, which is replaced by a significant increase in the susceptibility at low temperatures, reflects the presence of paramagnetic impurities and/or short segments of the chain. An analysis of the data in the

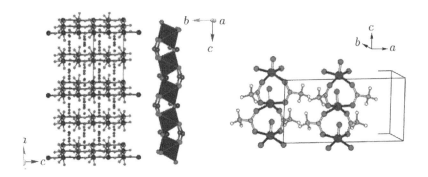

Fig. 2.3. One-dimensional chains of vanadium atoms in the structure of $VO(CH_3COO)_2$.

model of a quasi-one-dimensional Heisenberg linear chain with spin $S = 1/2$ gives an estimate of the exchange interaction $J \sim 215$ K in accordance with the Bonner–Fisher criterion (2.2).

First-principles theoretical calculations of the basic exchange interactions in vanadyl diacetate were carried out in [37]. Studies have shown that in the magnetic chain there are both superexchange spin interactions of V–O–V and supersuper exchange interactions through acetate groups along the V–O...O–V path, while the latter turn out to be dominant. The main exchange is antiferromagnetic, $J_{nn} \sim 203$ K and is realized via the carbon $2p_\pi$ orbital of the acetate ion $[CH_3COO]^-$ (right panel in Fig. 2.3).

An investigation of the temperature dependences of the magnetic susceptibility showed that, in analogy with the data of Ref. [36], the $\chi(T)$ curves exhibit a wide flat maximum, as shown in Fig. 2.4, which is replaced by a significant increase in susceptibility with decreasing temperature. Therefore, to analyze the temperature dependences of the magnetic susceptibility, we used the sum of two main contributions: the term χ_m, responsible for low-dimensional magnetism, and the Curie–Weiss term.

Various models were used to quantitatively analyze the contribution of χ_m. The best agreement was reached within the framework of the model of an antiferromagnetic Heisenberg chain with spin $S = 1/2$ [33] and the same model, but with allowance for the Dzyaloshinsky-Moriya (DM) interaction, the possibility of which follows from structural considerations for a non-centrosymmetric space group by analogy with one-dimensional magnets $[PM \cdot Cu(NO_3)_2 \cdot (H_2O)_2]_n$ (PM = pyrimidine) and Yb_4As_3 [38, 39] (solid lines in Figure 2.4).

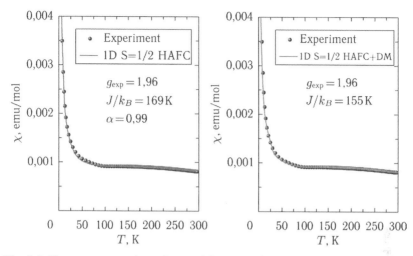

Fig. 2.4. The temperature dependence of the magnetic susceptibility of the compound VO(CH$_3$COO)$_2$ in the field $B = 0.3$ T (points) and the result of approximation in the framework of the AFM model of the Heisenberg chain with spin $S = 1/2$ (line) on the left panel and taking into account the interaction of DM line) in the right panel.

For Heisenberg antiferromagnetic chains with spin $S = 1/2$ with the DM interaction, the expression for the magnetic susceptibility was refined in [40]:

$$\chi(T) = g^2 \left(\frac{N_A \mu_B}{4k_B} \right) F\left(\frac{J}{k_B T} \right) \frac{1}{T}, \tag{2.5}$$

where for $x = J/k_B T$, $F(x)$ is an empirical rational function:

$$F(x) = \frac{1 + 0,08516x + 0,23351x^2}{1 + 0,73382x + 0,13696x^2 + 0,535\,368x^3}. \tag{2.6}$$

In our calculations, the value of the g-factor (see below) experimentally determined by the method of electron spin resonance (ESR), which amounted to $g = 1.96$ and was practically independent of temperature, was used.

The parameters obtained as a result of approximation in the framework of the two above-mentioned models are presented in Table 2.1. Using the values for the Curie constant, we can estimate the Curie–Weiss contribution at a level of 6–8%. By subtracting this contribution from the general dependence, one can single out the contribution of the low-dimensional magnetic

Table 2.1. The approximation parameters $\chi(T)$ in the framework of 2 models

Model	J/k_B [K]	$\chi_0 \cdot 10^{-4}$ [emu/ mol]	C [emu · K/ mol]	Θ [K]
The AFM spin chain of the Heisenberg spins $S = 1/2$	168.8	1.81	0.0205	−0.53
The AFM Heisenberg spin chain $S = 1/2$ with allowance for the DM interaction	154.3	1.10	0.0192	−0.38

subsystem and estimate the temperature of the maximum of the magnetic susceptibility T_{max} (Fig. 2.5). When the Heisenberg chain with spin $S = 1/2$ $T_{max} \sim 216$ K is approximated within the framework of the AFM model we have $T_{max} \sim 216$ K. Taking into account the interaction of DM within the same model we have, $T_{max} \sim 201$ K. The scale of the exchange interactions inside the chain is quite large and amounts to $J = -169$ K and $J = -154$ K, for the AFM and AFM+DM models, respectively The corresponding Bonner–Fisher criterion estimates for the two models were:

$$\frac{\chi_{max} |J|}{N_A g^2 \mu_B^2} = 0.0717, \quad \frac{k_B T_{max}}{|J|} \approx 1.278$$

and

$$\frac{\chi_{max} |J|}{N_A g^2 \mu_B^2} = 0.0736, \quad \frac{k_B T_{max}}{|J|} \approx 1.305$$

respectively, which is in reasonable agreement with the theoretical values of 0.07346 and 1.282 [33]. As can be seen from Fig. 2.5, there is a noticeable difference in the values of T_{max} on the $\chi_m(T)$ dependences. Therefore, to more correctly isolate the contribution of the one-dimensional chain and the corresponding estimates of the exchange integrals, ESR spectroscopy is used in this paper.

The ESR spectra of vanadyl diacetate were measured over a wide temperature range of 6–500 K. Evolution of the ESR spectra with a temperature variation is shown in Fig. 2.6. An analysis of the results obtained showed that the experimental ESR spectra are not described by a single line, but represent a superposition of two Lorentzian

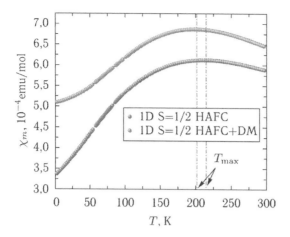

Fig. 2.5. Temperature dependences of the magnetic susceptibility of the compound $VO(CH_3COO)_2$ at $B = 0.3$ T, after subtracting the Curie–Weiss contribution, obtained by two methods: by approximation in the framework of the AFM model of the Heisenberg chain with spin $S = 1/2$ (lower curve) and the same model taking into account the DM interaction (upper curve).

components that correspond to the contributions of one-dimensional chains and paramagnetic impurities and/or chain breaks.

The results of the approximation are shown in Fig. 2.6, and an example of an expansion of the absorption line into spectral components is shown in Fig. 2.7. In the high-temperature region up to ~80 K, the main contribution to the absorption is made by the component L_1, then the integral intensities of both components are compared, and with further cooling of the sample, the component L_2 predominates.

The temperature dependences of the ESR parameters for two resolved spectral components (Fig. 2.8) exhibit essentially different behaviour. The effective g-factor for L_1 is $g_1 = 1.962$ and is practically independent of temperature. This value is characteristic for vanadium ions in the state V^{4+} ($S = 1/2$) [41, 42]. At the same time, the effective g-factor for L_2 increases significantly (from 1.937 to 1.963) with a decrease in temperature over the entire measured range. The width of the ESR line for both components varies nonmonotonically: it first decreases with cooling, passes through a minimum, and then rapidly increases. Different nature of the lines L_1 and L_2 is manifested in the temperature dependence of the integral ESR intensity $\chi_{ESR}(T)$, which is proportional to the concentration of paramagnetic centres in the sample (Fig. 2.8 *b*). It can be seen that the integral ESR intensities (the contributions to the

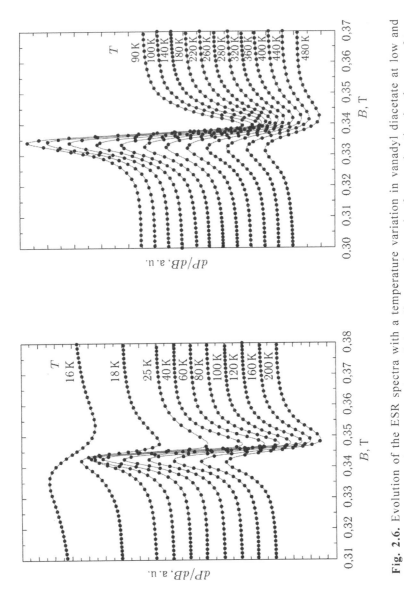

Fig. 2.6. Evolution of the ESR spectra with a temperature variation in vanadyl diacetate at low and high temperatures. Points are experimental data, lines are the result of approximation by the sum of two Lorentzians.

magnetic susceptibility) behave differently for the two components when the temperature is varied. The integrated ESR intensity of the L_1 line demonstrates the low-dimensional behaviour over the entire temperature range. On the contrary, a characteristic Curie–Weiss curve is observed for the L_2 line. The total dependence is analogous to the temperature dependence of the magnetic susceptibility obtained from measurements on a SQUID magnetometer.

The temperature dependence of the integrated ESR intensity for the L_1 line, which corresponds to a low-dimensional contribution to the magnetic susceptibility, was approximated by several models for low-dimensional systems, including models of isolated and interacting dimers and a one-dimensional antiferromagnetic Heisenberg chain, including the interaction of DM. In the calculations, the experimentally determined g-factor $g_1 = 1.96$ was used. The parameters obtained as a result of approximation in the framework of various models are collected in Table 2.2. It turned out that analysis within the framework of dimer models leads to unsatisfactory results for both the Heisenberg model and the Ising model. A good agreement with the experimental data, similar to the data on the analysis of the static magnetic susceptibility, was obtained within the framework of the model of an antiferromagnetic Heisenberg chain with spin $S = 1/2$ and a model with allowance for the DM-interaction. And, as can be seen from Fig. 2.8 *b*, the first model proves to be preferable. Despite a somewhat worse description within the framework of the model, taking into account the DM-interaction, the values of the exchange integral obtained in two different ways practically coincide, amounting to $J \sim 167$ K. The estimates of the exchange integral for both models used at $T_{max} = 216$ K for the AFM of the Heisenberg chain with spin $S = 1/2$ were:

$$\frac{\chi_{max}|J|}{N_A g^2 \mu_B^2} = 0.0716 \text{ and } \frac{k_B T_{max}}{|J|} \approx 1.29,$$

and for the same model, taking into account the interaction of the DM:

$$\frac{\chi_{max}|J|}{N_A g^2 \mu_B^2} = 0.072 \text{ and } \frac{k_B T_{max}}{|J|} \approx 1.285,$$

that is in good agreement with the Bonner–Fisher criteria (2.2).

It is interesting to note that the behaviour of $\Delta B_1(T)$ observed by us is similar to that obtained earlier for one-dimensional antiferromagnets $BaCu_2Ge_2O_7$, $KCuF_3$ [43]. In [43], the temperature dependences of the width of the ESR line was described using the Oshikawa–Affleck model for Heisenberg antiferromagnetic chains $S = 1/2$ with the DM-interaction [44–47], which is considered valid in the temperature range up to $1/2T_{max}$.

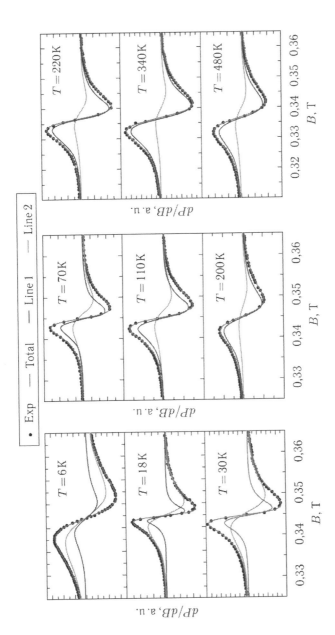

Fig. 2.7. The first derivatives of the ESR absorption spectrum for vanadyl diacetate at different temperatures (points). the red lines are the approximation of the spectrum by the sum of two Lorentzians. the blue and green are the resolvable components of the spectrum.

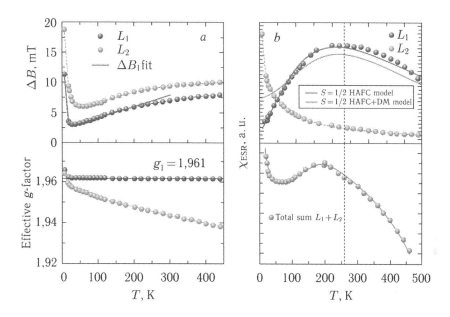

Fig. 2.8. Temperature dependences of the width of the ESR line, the effective g-factor (*a*), and the integrated ESR intensity for individual components and their sums, normalized to an appropriate constant in comparison with the magnetic susceptibility measurements on a SQUID magnetometer (*b*). The solid lines on the upper part of the right panel show the result of approximation in the framework of the Heisenberg chain models and the chain model, taking into account the DM-interaction.

Table 2.2. The exchange parameters obtained in the framework of different models from the ESR data

Model of the magnetic structure	J, K	g
1D infinite AFM Heisenberg spin chain $S = 1/2$	167 ± 3	1.962
1D infinite AFM Ising spin chain $S = 1/2$	226 ± 11	1.962 (given)
Dimer Heisenber model with spin $S = 1/2$	143 ± 5	$0.73 \pm 0{,}02$
Dimer Ising model with spin $S = 1/2$	238 ± 5	$1.23 \pm 0{,}05$
1D infinite AFM Heisenberg spin chain $S = 1/2$, taking into account the interaction of the DM	168 ± 3	1.962 (given)

On the upper panel of Fig. 2.8 *a* is the result of the approximation of the temperature dependence of the width of the ESR line in the temperature range 6–200 K in the framework of the Oshikawa–Affleck model in accordance with expression

$$B(T) = AT + B(T - T) \ , \tag{2.7}$$

where A and B are the coefficients, T_N is the Néel temperature, and n is the exponent. The linear term in this expression was predicted by the Oshikawa–Affleck theory. According to this theory, the width of the ESR line for Heisenberg antiferromagnetic chains with spin 1/2 should be directly proportional to the temperature at $T \ll J$. The second term becomes significant at low temperatures and is interpreted as the appearance of three-dimensional antiferromagnetic fluctuations as the Néel temperature approaches [43].

As can be seen from Fig. 2.8 *a*, when the temperature is lowered below $T_{max}/2$, the half-width of the line follows the linear dependence of $\Delta B \sim T$ over a wide range of temperatures, and at low temperatures the contribution of another component ΔB, which can be called ΔB_{3D}, becomes significant. The values of the model parameters obtained from the approximation were $T_N = 0.23$ K and $n \approx -1.17$, the latter being in satisfactory agreement with the value $n = -1.1$ obtained for the one-dimensional antiferromagnet $BaCu_2Ge_2O_7$ [43].

The best agreement with the experimental data is obtained for the model of a homogeneous antiferromagnetic Heisenberg chain $S = 1/2$. The exchange interaction parameter in the chain is determined from ESR data as $J = 167$ K. According to the estimate from $T_N = \sqrt{J \times J'}$, the parameter of the interchain exchange interaction J' in vanadyl diacetate is $J' < 0.02$ K. Thus, the ratio of the exchange interaction between the chains and within the chain reaches an extremely small value $J'/J \sim 10^{-4}$.

2.2. A homogeneous chain with competing interactions

Some complication of the model of a homogeneous Heisenberg chain, where neighbouring magnetic centers are bound by the J_1 exchange interaction, can be done by adding an interaction with the next nearest neighbor J_2, as shown in the inset to Fig. 2.9. This can lead to frustration and the emergence of new basic states. The Hamiltonian of such a system can be written as:

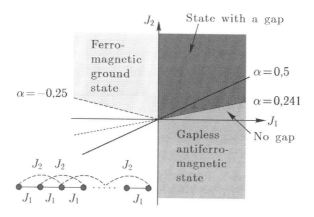

Fig. 2.9. Phase diagram of a chain Heisenberg magnet with competing exchanges [48]. The inset shows an antiferromagnetic chain with the exchange between the nearest neighbors J_1 and with an exchange with the next nearest neighbor J_2.

$$\hat{H} = \sum_i \left(J_1 S_i S_{i+1} + J_2 S_i S_{i+2} \right). \tag{2.8}$$

In such a record, $J > 0$ corresponds to an antiferromagnetic interaction, $J < 0$ – to a ferromagnetic interaction. The phase diagram of a chain composed of spins $S = 1/2$ is shown in Fig. 2.9.

For $J_1 > 0$ and $J_2 > 0$ (competing antiferromagnetic exchanges), the ground state of such a system is the spin liquid. An increase in the ratio of the exchange parameters $\alpha = J_2/J_1$ induces a phase transition from the gapless to the spin-gap state of the dimerized phase. The critical point α_c is determined numerically and is 0.241. With a further increase in the ratio of the exchange parameters to $\alpha = 0.5$, the so-called Majumdar–Ghosh point is realized when the ground state is a set of singlet pairs composed of nearest neighbors.

At $J_1 < 0$ and $J_2 > 0$ (ferromagnetic and antiferromagnetic exchanges) and $-1/4 \leq \alpha \leq 0$, the ground state is ferromagnetic, and for $\alpha < -1/4$ it is an incommensurate helix with a total moment $S = 0$ and angle of rotation $\cos \varphi = -J_1/4J_2 = 1/4\alpha$. It is assumed that in this incommensurable state the gap is suppressed [48–52].

To date, a large number of metal oxide compounds with homogeneous chains with antiferromagnetic exchange J_1 have been detected in which it is necessary to take into account the frustrating antiferromagnetic exchange with a next-nearest neighbour J_2. In [53] an attempt was made to systematize such chains with competing interactions in copper-based compounds as they approach

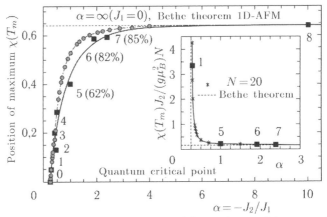

Fig. 2.10. Dependence of the reduced temperature of the maximum of the susceptibility T_m/J_2 on the parameter α for alternating chains in cuprates (black squares). 0: $\alpha_c = 1/4$; 1: Li_2CuZrO_4; 2: $Pb_2[CuSO_4(OH)_2]$; 3: $Rb_2Cu_2Mo_3O_{12}$; 4: $Cs_2Cu_2Mo_3O_{12}$; 5: $LiCu_2O_2$; 6: $NaCu_2O_2$; 7: $LiCuVO_4$; 8: $SrCuO_2$. The values of the angle of rotation in the helix are given in parentheses. The small points are the result of the calculation by the method of complete diagonalization in model J_1–J_2 on the cluster $N = 20$. On the inset: the dependence of the reduced susceptibility at the maximum on α [53].

the quantum critical point $\alpha_c = -1/4$. As shown in Fig. 2.10, as the parameters approach this point, the angle of rotation of neighbouring spins in the spiral decreases, thereby increasing its period.

2.2.1. Lithium–copper zirconate

In the diagram shown (Fig. 2.10), the γ-Li_2CuZrO_4 metal oxide is located closest to the quantum critical point. Presumably, the ground state in it is an incommensurable spiral. Figure 2.11 shows a fragment of a crystal structure in which copper ions form ribbons of edge-shared CuO_4 squares. The $\angle$–Cu–O–Cu bond angle is 94°, which corresponds to ferromagnetic exchange and is quite close to the critical value of the 96° bond angle separating the ferro- from the antiferromagnetic exchange interactions [54]. Lithium ions occupy two positions, the position of Li1 being between different layers of Cu–Li, half-filled and mobile.

To establish the temperature of the magnetic ordering in the Li_2CuZrO_4 system, detailed low-temperature measurements of the magnetic susceptibility, specific heat, and muon scattering spectra, were made and are presented in Fig. 2.12 [55]. It can be seen that the maximum at 8 K on the temperature dependence of the magnetic susceptibility $\chi(T)$ does not correspond to the formation of

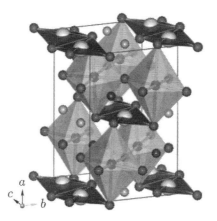

Fig. 2.11. Elementary cell γ-Li$_2$CuZrO$_4$. Copper Cu^{2+} cations are in a square environment of oxygen atoms, Zr^{4+} cations are in an octahedral oxygen environment, isolated spheres denote two positions of Li$^+$.

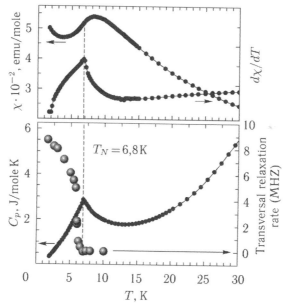

Fig. 2.12. Upper panel: the temperature dependence of the magnetic susceptibility and its derivative in Li$_2$CuZrO$_4$. Bottom panel: temperature dependences of specific heat and transverse muon relaxation rate in Li$_2$CuZrO$_4$ [55].

a magnetically ordered state. Whereas the long-range magnetic order arises at a lower temperature $T_N = 6.8$ K, at which the temperature anomaly is demonstrated the dependence of the derivative of the magnetic susceptibility $d\chi/dT$, specific heat C_p, and the transverse relaxation rate of muons.

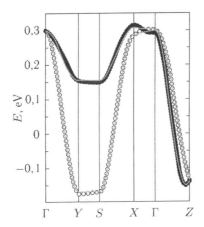

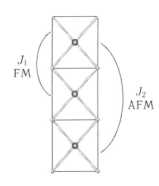

Fig. 2.13. The band structure of the Li_2CuZrO_4 metal oxide calculated by LSDA + FPLO [53].

Fig. 2.14. Schematic representation of a copper chain in Li_2CuZrO_4.

To study the quantum ground state of Li_2CuZrO_4, the energy spectrum was calculated by the method of local spin approximation (LSDA + FPLO, Fig. 2.13) and the exchange magnetic interactions shown in Fig. 2.14.

The main exchange magnetic interaction in the chain between the nearest neighbours was found ferromagnetic: $-J_1 = 151$ K, next-nearest neighbours – antiferromagnetic: $J_2 = 35$ K ($\alpha = -J_2/J_1 = 0.23$), which puts the Li_2CuZrO_4 system near the quantum the critical point $\alpha_C = 0.25$ separating the ferromagnetic chain $0 < \alpha < \alpha_C$ and the antiferromagnetic helicoid $\alpha > \alpha_C$. Interchain interaction was $J_3 = 6$ K [53]. The proposed theoretical model describes quite well the nuclear magnetic resonance data on ^7Li, which assume a difference in the angles between neighbouring spins $\Delta\varphi = 33°$, which corresponds to the value of the parameter $\alpha = 0.33$ [56]. It should be noted, however, that the neutron diffraction data performed on a Li_2CuZrO_4 powder sample gives an estimate of $\Delta\varphi = 88°$ and $\alpha = 6.2$, which puts this compound far from the quantum critical point [57].

2.2.2. Isostructural cuprates of lithium and sodium

The $LiCu_2O_2$ and $NaCu_2O_2$ crystals have orthorhombic symmetry, the space group *Pnma* [59, 60]. The parameters of the crystal lattice turned out to be related by the relation $a \approx 2b$, which leads to twinning with crystal growth and difficulties in their orientation [60–64]. Since the ionic radii of Li^+, Cu^+ and Cu^{2+} are close, it may lead to lithium substitution of copper positions in lithium cuprate, in contrast to sodium cuprate, since the ionic radius of Na^+ is much larger. When studying the properties of $LiCu_2O_2$ and $NaCu_2O_2$, the quality of single crystals and the 'intrinsic' chemical disorder play a key role and lead to a substantial variation of the experimental data.

Interest in these compounds was due to the presence of two copper cations in the structure of the oxidation state +1 and +2 in a 1:1 ratio. Such a structure can be represented as a successive alternation of three layers: $-Li/Na$ O $Cu^{2+}O-$, $-Cu^+$ and $-O$ Cu^{2+} O Li/Na− (Fig. 2.15, left panel). Along the *a* axis, there are chains of alternating Cu–O− and Li/Na–O–pyramids, and along the *b* axis there are linear chains consisting only of Cu–O− or Li/Na–O–pyramids. The magnetic structure is formed by two elongated along the axis *b* exchange-linked linear chains of cations Cu^{2+}, which belong to two adjacent $Li(Na)CuO_2$-layers. Cu^{2+} cations lie in the base of CuO_5 pyramids, connected in a chain through a common face. Between each other, the chains of the pyramids are connected through a common side face so that the top of the pyramid of one layer lies at the bottom of the pyramid of the other layer (Fig. 2.15, right panel). The angle at the base of the pyramid between Cu^{2+}–O–Cu^{2+} ions is 87.2° in lithium cuprate [60] and 92.9° in sodium cuprate [64]. As a result, Cu^{2+} cations form a double chain on the spins $S = 1/2$. Such a configuration, depending on the magnitudes and signs of the exchange interactions, can be considered both as a zigzag chain and as a spin ladder.

Figure 2.16 shows the temperature dependence of the magnetic susceptibility in $LiCu_2O_2$ [66]. With decreasing temperature, a broad maximum is observed on the curve, which is characteristic of low-dimensional magnets, indicating the establishment of a regime of

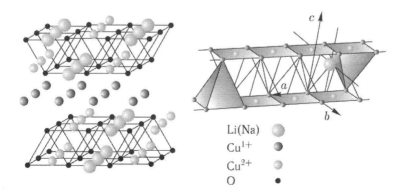

Li(Na)
Cu^{1+}
Cu^{2+}
O

Fig. 2.15. Left panel: crystal structure of $LiCu_2O_2$, $NaCu_2O_2$ compounds. Right panel: double chain based on Cu^{2+} ions [65].

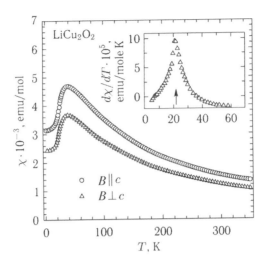

Fig. 2.16. The temperature dependence of the magnetic susceptibility of $LiCu_2O_2$ along the chains and perpendicular to them [66]. On the inset: the derivative of magnetic susceptibility.

short-range correlations within the ladder. When the temperature is lowered to $T_{N1} = 22$ K, a transition occurs in this compound in the antiferromagnetic state. A feature at 22 K is also observed in the temperature dependence of the specific heat shown in Fig. 2.17, which confirms the presence of a second-order phase transition associated with a change in the magnetic structure at T_{N1}. More detailed consideration of specific heat curve reveals another transition at $T_{N2} = 24$ K [66].

The temperature dependence of the magnetic susceptibility and the field dependence of the magnetization of $NaCu_2O_2$ single crystals were investigated in Ref. [15]. For this compound, a broad maximum at $T \sim 50$ K is also observed on $\chi(T)$. A sharp increase in $\chi(T)$ in the low-temperature region and a small bending of $M(B)$ in weak fields is attributed to the weak ferromagnetic interaction in the system [15]. However, this can also be caused by the presence of a small amount of ferromagnetic impurity.

Specific heat of $NaCu_2O_2$ single crystals was studied in [15, 65, 66]. The single anomaly in specific heat at $T_N = 13.4$ K was observed on $NaCu_2O_2$ single crystals [65], i.e., the formation of the basic magnetically ordered state in the sodium cuprate occurs in one stage, unlike lithium cuprate, where there are two closely spaced phase transitions.

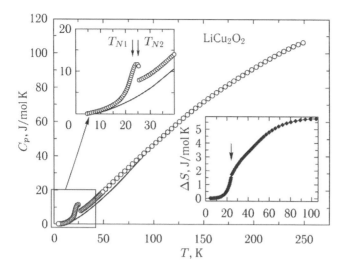

Fig. 2.17. The temperature dependence of specific heat of $LiCu_2O_2$ [66]. Upper insert: specific heat near the double phase transition; lower insert: magnetic entropy.

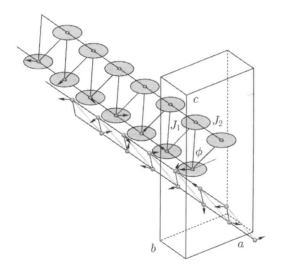

Fig. 2.18. The elementary cell $LiCu_2O_2$ (copper ions are shown by spheres) and the planar helicoidal magnetic structure at $T = 2$ K (the directions of the magnetic moments are shown by arrows) [66].

On the basis of neutron diffraction studies [66], it was established for lithium cuprate that each of the two spin chains aligns with the spiral structure shown in Fig. 2.18, in which the spins lie in the ab planes and coincide with each other by means of the vector $\mathbf{q} = (1/2, k, 0)$, where $k = 1 - \zeta$ plays the role of the parameter of incommensurability of the lattice period with the magnetic spiral period and depends on the temperature, as shown in Fig. 2.19. The magnitude of the incommensurability parameter for $LiCu_2O_2$, according to the estimate, is $\zeta = 0.174$.

For a more detailed study of the magnetic structure, samples with an isotope of ^7Li were synthesized, on which additional neutron diffraction studies were carried out [67] and experiments on nuclear magnetic resonance [68, 69]. Recent studies have confirmed that at low temperatures the shape of the ^7Li NMR line is typical for an incommensurable static modulation of the magnetic field corresponding to the spiral structure of the Cu^{2+} magnetic moments.

The appearance of the helicoidal spin configuration can be explained by the competition between the antiferromagnetic interaction J_1 between the next-nearest neighbours in the chain and the ferromagnetic interaction J_2 between the neighbours following the

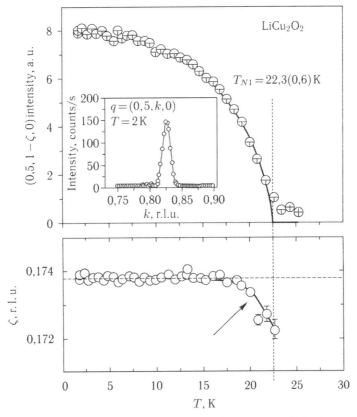

Fig. 2.19. The upper panel: the temperature dependence of the intensity of the magnetic peak (0.5, 0.826; 0). In the inset: k-scanning along this reflex at $T = 2$ K. The lower panel: the temperature dependence of the parameter ζ [66].

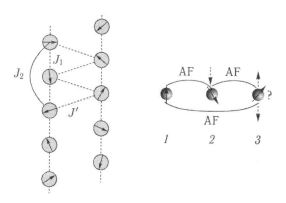

Fig. 2.20. Left panel: the simplest scheme of interactions in the spin ladder in Li(Na)Cu$_2$O$_2$ [70]; right panel: the formation of a non-collinear order in the chain in the competition of antiferromagnetic exchanges.

nearest (Fig.2.20, left panel). It is also necessary to take into account the interaction with even more distant neighbors (J'). As a result, the magnetic moments in neighbouring planes (or in neighboring chains) are rotated by a constant angle. In addition, the formation of a spiral or non-collinear ordering in a chain is possible if both exchange integrals J_1 and J_2 are antiferromagnetic. In the right panel of Fig. 2.20 a frustration scheme is presented on the example of three magnetic atoms in a chain with antiferromagnetic interaction. Analogous to the trimers considered earlier, antiferromagnetic interactions between ions 1–3 and 2–3 compete with each other, as a result of which non-collinear ordering is realized.

Neutron diffraction studies of sodium cuprate were made in [69, 70]. In NaCu$_2$O$_2$, like in the lithium cuprate, magnetic moments form a spiral magnetic structure that is incommensurate with the period of the crystal lattice. The parameter of incommensurability in NaCu$_2$O$_2$ is $\zeta = 0.227$. The magnitude of the angle of inclination of the magnetic moments in the helicoid is $\varphi_0 = 81.7°$ to the c axis [70]. The main exchange integrals in NaCu$_2$O$_2$ are estimated as – $J_1 = 16.4$ K, $J_2 = 90$ K, $J' \sim 7$ K. It turned out that the strongest interaction J_2 does not work between the nearest neighbours in the chain, but between the next-nearest neighbours. This is due to the fact that the Cu–O–Cu bond angle is equal to $92.9°$ and is close to the critical value of $\sim 94°$, at which the sign of the super-exchange changes. Thus, the value of J_1 is small.

For the LiCu$_2$O$_2$ compound, preliminary calculations corresponding to a simple model with three exchange integrals (the left panel of

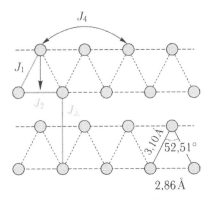

Fig. 2.21. Schematic representation of the exchange interaction in the $LiCu_2O_2$ compound [67], the Hamiltonian (2.9).

Fig. 2.20) did not give a good agreement with the experimental data, and a more complicated scheme of exchange interaction with four exchange integrals, shown in Fig. 2.21. The Hamiltonian of this system has the form:

$$\hat{H} = \sum_{ij}(J_1\hat{S}_{i,j}\hat{S}_{i+1,j} + J_2\hat{S}_{i,j}\hat{S}_{i+2,j} + J_4\hat{S}_{i,j}\hat{S}_{i+4,j} + J_{\perp}\hat{S}_{i,j}\hat{S}_{i,j+1}), \qquad (2.9)$$

where the index i numbers the spind in both zigzag chains, and the index j – is the chains themselves. On the basis of data on neutron diffraction, a number of models have been proposed leading to spiral ordering, and calculations have also been made in the local density of states approximation (LDA) presented in Table 2.3 [67].

Each of these models describes well the behaviour of the magnetic susceptibility, however, in [67], an adequate choice for inelastic neutron scattering for small **k** was chosen in favour of Model 1.

Table 2.3. The values of exchange integrals in $LiCu_2O_2$

Exchange integrals	Model 1	Model 2	Model 3	LDA
J_1, meV	3.2 (0.5)	52.8 (4.0)	0 (given)	0.4
J_2, meV	−5.95	16.9	−7.0	−8.1
J_4, meV	3.7 (0.3)	−0.8 (0.1)	3.75 (0.05)	14.4
$J_{\perp}$, meV	0.9 (0.1)	0.12 (0.01)	3.4 (0.2)	5.7

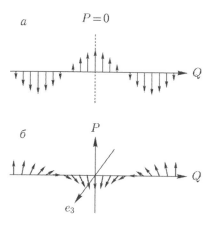

Fig. 2.22. Top panel: a sinusoidal spin density wave does not lead to polarization; bottom panel: a helicoidal spin density wave induces the appearance of polarization **P**, perpendicular to the vectors **Q** and $\mathbf{e}_3$ [76].

Thus, the formation of a spiral in $LiCu_2O_2$ is due mainly to the competition of the main exchange J_2 neighbour in a chain) and two interactions of another sign J_1 and J_4, as shown in Fig. 2.21.

The evolution of the magnetic helix in $LiCu_2O_2$ in the intermediate region $T_{N1} < T < T_{N2}$ was studied by synchrotron radiation in [71]. It was found that when the temperature is lowered, a sinusoidal spin structure is first formed that exists in the interval $T_{N1} < T < T_{N2}$, and then it is rearranged into a helicoidal structure at $T < T_{N1}$. The presence of two phase transitions associated with a change in the spiral magnetic structure in $LiCu_2O_2$ stimulated interest in the study of the permittivity and polarization in this compound [71–75]. It was shown in Ref. [76] that the formation of a helicoidal magnetic structure (spin-density wave) can induce the appearance of an electric polarization vector **P** that will be directed perpendicularly to the propagation vector of the helicoid **Q** and to the rotation vector of magnetic moments $\mathbf{e}_3$, as shown in Fig. 2.22 [76]. The model with induced polarization made it possible to describe multiferroics such as $TbMnO_3$, $DyMnO_3$ and $TbMn_2O_5$ [77, 78].

Figure 2.23 shows the temperature dependences of the permittivity ε and the polarization of P_c in $LiCu_2O_2$ [75]. The dielectric constant shows two anomalies for T_{N1} and T_{N2}, and the polarization $P \| c$ appears at the low-temperature phase transition T_{N1}. The results obtained indicate that $LiCu_2O_2$ is a multiferroic and can serve as a prototype of a ferroelectric with a one-dimensional helical magnetic structure.

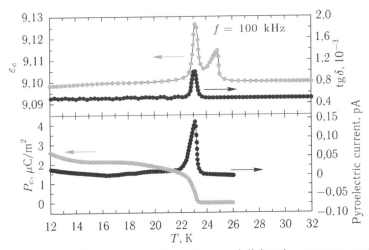

Fig. 2.23. Upper panel: temperature dependences of dielectric constant ε and loss tangent in $LiCu_2O_2$; lower panel: temperature dependences of the polarization along the c axis and pyroelectric current [75].

2.2.3. Rubidium–copper–aluminium phosphate

The crystal structure of rubidium–copper–aluminum phosphate $RbCuAl(PO_4)_2$ is monoclinic and has a space group $P2_1/c$ [79]. The non-centrosymmetric structure contains layers of zigzag chains of highly distorted CuO_6 edge-shared octahedra, as shown in Fig. 2.24. The chains are separated from each other by phosphorus tetrahedra PO_4 and aluminum bipyramids AlO_5. In this case, open channels are formed in the unit cell parallel to the axes a and c, in which the Rb^+ ions are located. The octahedrons of CuO_6 are strongly distorted in accordance with the Jahn–Teller effect, which is characteristic of copper ions Cu^{2+} (d^9).

The temperature dependence of the static magnetic susceptibility of $RbCuAl(PO_4)_2$, as shown in the upper panel of Fig. 2.25, at high temperatures obeys the Curie–Weiss law with the parameters $\chi_0 = 5.84 \times 10^{-4}$ emu/mol, $C = 0.435$ emu/mol K and $\Theta = 15$ K. As the temperature is lowered, the curve $\chi(T)$ exhibits a bend at $T = 10.5$ K, indicating the establishment of a long-range magnetic order, and passes through a maximum of the Schottky type [79]. The behaviour of $\chi(T)$ in the cooling regime in the field indicates the presence of a weak ferromagnetic component in the magnetic response of $RbCuAl(PO_4)_2$.

The effective magnetic moment, calculated from the value of the Curie constant, is $\mu_{eff} \approx 1.87\mu_B$ per formula unit and is in good

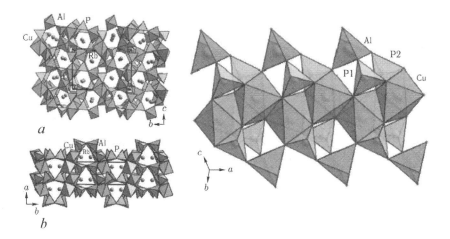

Fig. 2.24. General view of the crystal structure of RbCuAl(PO$_4$)$_2$ in different projections (on the left) and zigzag quasi-one dimensional chains of copper octahedra CuO$_6$ edge-shared in the magnetically active layers in the *ac* plane (on the right).

agreement with the expected $\mu_{\text{theor}} = ng^2 S(S+1)\mu^2$ per Cu^{2+} ion ($S = 1/2$) for $g = 2.14$, estimated from the ESR data. The positive sign of the Weiss temperature indicates the dominance of ferromagnetic correlations at high temperatures. This is also seen on the bottom of the top panel of Fig. 2.25 where the temperature dependence of the quantity $(\chi - \chi_0)$ $(T - \Theta)$ is presented. The horizontal line on the insert at high temperatures corresponds to the region of fulfillment of the Curie–Weiss law. When the temperature drops below ~ 120 K, the experimental curve deviates significantly from the horizontal line down, indicating an increase in the role of antiferromagnetic correlations of the short-range order.

As shown in the top panel of Fig. 2.25, at $T < 10.5$ K, the static magnetic susceptibility, measured in the ZFC and FC regimes, is strongly divergent. In weak magnetic fields, anomalies of the Schottky type are observed in the magnetic susceptibility and specific heat. The specific heat of $C_p(T)$ shows a peak at $T_N = 10.5$ K, shown in the lower panel of Fig. 2.25, which can be associated with the establishment of the long-range magnetic order. The amplitude of the peak is suppressed because most of the entropy in the low-dimensional RbCuAl(PO$_4$)$_2$ system is released substantially above the Néel temperature. A brighter anomaly, the nature of which is not clear, is clearly visible at $T_{\text{max}} \sim 3.5$ K.

The magnetization isotherms, measured at different temperatures, are shown in Fig. 2.26. It can be seen that $M(B)$ practically reaches

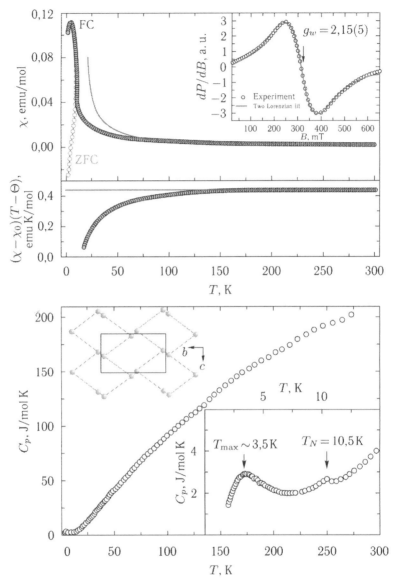

Fig. 2.25. The upper panel: temperature dependences of the static magnetic susceptibility of $RbCuAl(PO_4)_2$, measured in the ZFC and FC regimes at $B = 0.1$ T. On the inset: ESR spectrum at 300 K. On the bottom part: temperature dependence of $(\chi-\chi_0)$ $(T-\Theta)$. The lower panel: temperature dependence of the specific heat. Bottom inset: enlarged low-temperature region. Top inset: the magnetic subsystem topology.

full saturation (theoretically expected $M_S = gS\mu_B = 1.1\mu_B$/f.u.) at $B \sim 9$ T at 2 K. Saturation in moderate fields indicates the coexistence of ferromagnetic and antiferromagnetic exchanges in

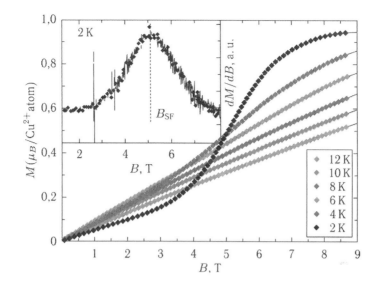

Fig. 2.26. Isotherms of magnetization $M(B)$ for RbCuAl(PO$_4$)$_2$ at different temperatures. On the inset: the first derivative of the field $(dM/dB)(B)$.

the system under investigation and allows to assume a non-trivial spin-configuration structure. In addition, the anomaly induced by the magnetic field at $B \sim 5$ T is present on the $M(B)$ dependences, which is probably related to the spin-reorientation (spin-flop) transition. The position of this transition is slightly shifted toward smaller fields with increasing temperature and it is not observed at $T > T_N$.

Evolution of the ESR spectra measured on a powder sample of RbCuAl(PO$_4$)$_2$ is shown in Fig. 2.27 *a*. The best description of the shape of the line can be achieved by the sum of two Lorentz-type lines. In the entire investigated temperature range, the L_1 mode predominates, and the narrow and low-intensity mode L_2 corresponds to the presence of ~2% of the paramagnetic impurity. The temperature dependences of the main ESR parameters are shown in Fig. 2.27 *b*. The integrated intensity of L_1 χ_{ESR} demonstrates a monotonic growth with decreasing temperature and its behaviour is consistent with the temperature dependence of the magnetic susceptibility. The effective *g*-factor and the width of the absorption line are practically independent of temperature at $T > 35$ K. In the paramagnetic region the ESR absorption is characterized by a *g*-factor of $g = 2.14$, typical for copper ions Cu^{2+} ($S = 1/2$) in a tetragonal-distorted octahedral coordination. In the vicinity of $T \sim 10$ K, the ESR signal is degraded and, ultimately, disappears, indicating the opening of the gap in the

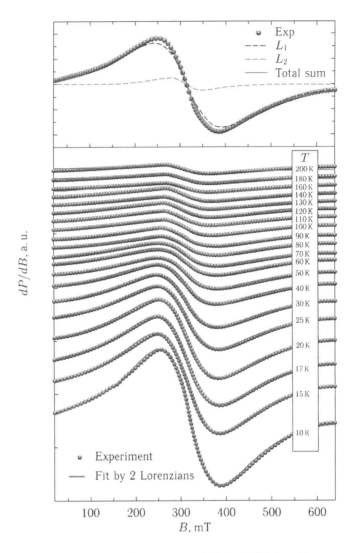

Fig. 2.27 a. Evolution of the ESR spectra of RbCuAl(PO$_4$)$_2$ with a temperature variation and an example of the expansion of the spectrum into two components at $T = 20$ K (at the top).

spectrum of magnetic excitations when the long-range magnetic order is established in accordance with the static magnetic data.

The first-principles calculations were performed to obtain a microscopic picture of the exchange interactions. The main ways of exchange interactions J_1–J_4 in RbCuAl(PO$_4$)$_2$ are shown in Fig. 2.28. It has been established that intra-chain exchanges J_1, J_2 and J_3, shown in Fig. 2.28 and listed in Table 2.4 dominate in the system. Of these,

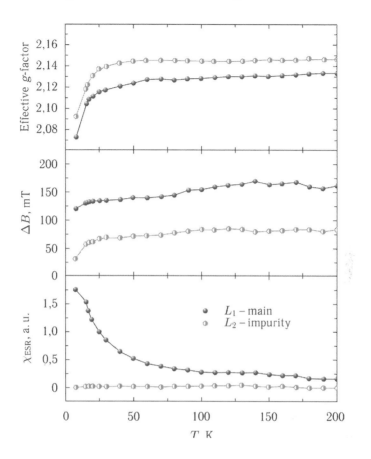

Fig. 2.27 *b*. Temperature dependences of the main parameters for two resolved ESR lines in RbCuAl(PO$_4$)$_2$.

J_1 and J_3 follow a supersuper exchange mechanism, and J_2 follow a superexchange mechanism. The magnetic structure is alternating zigzag APM–AFM chains of type J and αJ (where $\alpha \approx 0.4$), with the exchange between the next-nearest neighbours being ferromagnetic in accordance with the coexistence of the antiferro- and ferromagnetic components in the magnetic response of the system under study. Ferromagnetic orientation of the spins along the J_3 path removes frustration and allows the system to go over into an ordered state. Interchain exchange interactions J_4 are antiferromagnetic and small in accordance with large interatomic distances (Cu...Cu = 7.48 Å).

Thus, the magnetic subsystem in RbCuAl(PO$_4$)$_2$ is represented by chains of edge-shared octahedra combined with weak interchain interactions through polyhedra PO$_4$ and AlO$_5$. Since chains of CuO$_6$ octahedra alternate through *cis* and *trans* bonds, exchanges of both

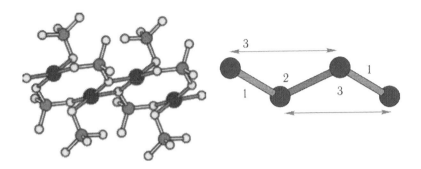

Fig. 2.28. Left panel: $Cu(PO_4)$ chains in the structure of $RbCuAl(PO_4)_2$. Right panel: the main paths of intrachain exchange interactions J_1, J_2 and J_3 in $Cu(PO_4)$.

Table 2.4. The parameters of exchange interactions in the structure of $RbCuAl(PO_4)_2$ (the plus sign corresponds to the FM interaction, the minus – to AFM)

U, eV	3	4	5
J_1, K	−47	−38	−31
J_2, K	−147	−108	−77
J_3, K	10	8.6	7.3
J_4, K	−0.32	−0.26	−0.20

antiferro- and ferromagnetic types occur, similar to the situation observed in pyroxenes [80, 81]. The exchange between the chains is antiferromagnetic in agreement with the establishment of a long-range AFM order and the deviation of the experimental data from the Curie–Weiss law, as follows from the data presented at the bottom of the upper panel of Fig. 2.25.

2.2.4. Cesium–copper vanadium-diphosphate

Cesium-copper vanadium-diphosphate $Cs_2Cu(VO)(P_2O_7)_2$ is characterized by a non-centrosymmetric orthorhombic crystal structure, the space group $Pn2_1a$ [82]. The structure is formed by Cu ions in a square-pyramidal and planar-square environment, tetragonal pyramids VO_5, and diphosphate groups P_2O_7 bound as Cu–O–V, Cu–O–P and V–O–P. These groups form open channels for cesium ions, as shown in Fig. 2.29 *a*. In the unit cell there are three crystallographically independent positions for the magnetic ions of copper and vanadium. Cu1 and Cu2 slightly emerge from the basal plane of the square oxygen pyramids, and Cu3 ions are in

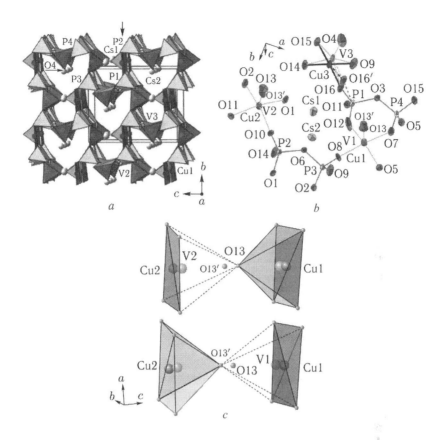

Fig. 2.29. The crystal structure of $Cs_2Cu(VO)(P_2O_7)_2$: a polyhedral view with channels for cesium in the [100] direction (*a*); basic structural elements (*b*); two possible configurations in the arrangement of ions Cu, V and O (*c*).

a planar-square environment, as shown in Fig. 2.29 *b*. In turn, the magnetic ions V^{4+} are coordinated by five oxygen ions into a square pyramid with one shortest distance V–O, which is typical for vanadyl groups. In this case, the V^{4+} ions are displaced from the basal plane in the direction of the pyramid vertex and form dimers with Cu^{2+} ions, as shown in Fig. 2.29 *b*, *c*. The corner-shared square pyramids of vanadium and copper and diphosphate groups form honeycomb layers parallel to the *ac* plane.

Evolution of the ESR absorption spectra for a powder sample of $Cs_2Cu(VO)(P_2O_7)_2$ (more precisely, $Cs_2Cu_{1,1}(VO)_{1,9}(P_2O_7)_2$) with a temperature variation is shown in Fig. 2.30 *a*. The spectra contain a single exchange-narrowed absorption line. Analysis of its shape, however, requires the presence of two different resonant modes, as shown in Fig. 2.30 *b*. The average effective *g*-factors at room temperature are g_1 = 2.001 and g_2 = 2.037. Since there are two magnetically active ions Cu^{2+} (d^9, $S = 1/2$) and V^{4+} (d^1, $S = 1/2$) in the system under study, it is natural to assign two observed modes to the responses from the copper and vanadium subsystems. Taking into account the presence of three non-equivalent positions for copper and vanadium pairs, Cu1/V1, Cu2/V2 and Cu3/V3, the ESR spectrum of such a concentrated magnetic system is a superposition of two statistically averaged modes corresponding to the response of the paramagnetic centres Cu^{2+} and V^{4+} connected by the exchange interaction, as shown in Fig. 2.29 *c*. If we neglect the effects of tetragonal and trigonal distortions, the crystal field splits the *d*-shell into a triplet Γ_5 and a doublet Γ_3. The corresponding expressions for the principal components of the anisotropic *g*-tensor can be written as [83]:

$$g_{\parallel} = g_{zz} = g_e - \frac{8\lambda}{\Delta},$$
$$g_{\perp} = g_{xx} = g_{yy} = g_e - \frac{2\lambda}{\Delta}, \tag{2.10}$$

or in the isotropic case:

$$g = g_{zz} = g_{xx} = g_{yy} = g_e - \frac{4\lambda}{\Delta}, \tag{2.11}$$

where g_e is the *g*-factor of the free electron, λ is the spin-orbit interaction constant, and Δ is the splitting energy by the crystal field.

Usually $\lambda > 0$ for ions d^1, but $\lambda < 0$ for ions d^9, and, as a consequence, we can expect $g < g_e$ and $g > g_e$ for ions d^1 and d^9, respectively. However, the values of the *g*-factors for the two resonant modes obtained from the analysis of the experimental data differ significantly from the typical values in compounds with Cu^{2+} (d^9, $S = 1/2$) and V^{4+} (d^1, $S = 1/2$). These values, as a rule, vary in the ranges $g = 1.93$–1.98 [84–89] and $g = 2.02$–2.22 [90–93] in the plane-square, pyramidal and octahedral oxygen coordination for V^{4+} and Cu^{2+}, respectively. If the value $g_2 \sim 2.04$ can still be attributed to the signal from Cu^{2+} ions, then an abnormal decrease in the negative shift of g_1 relative to g_e indicates an important role of exchange interactions between the V and Cu subsystems. In support of the hypothesis of a strong exchange interaction, the fact that both the fine and hyperfine structure expected in the presence of magnetic isotopes ^{63}Cu and ^{65}Cu ($I = 3/2$) or ^{51}V ($I = 7/2$) is not allowed in experimental spectra even at the lowest temperatures.

In the first approximation of the perturbation theory, the spin Hamiltonian for two different $S = 1/2$ spins i and j (with different g-factors) and the isotropic exchange interaction J can be written as [94]:

$$\hat{H} = \mu_B B_z \left(g_{iz} S_{iz} + g_{jz} S_{jz} \right) + J \left(S_i \cdot S_j \right), \qquad (2.12)$$

where B_z is the magnetic field along the z axis, which corresponds to the main anisotropy axis for both g-tensors, and μ_B is the Bohr magneton. In the absence of exchange interaction between the spins i and j (similar to the case of isolated ions), the transition arises when $h\nu = g_{iz} \mu_B B_z$ and $h\nu = g_{jz} \mu_B B_z$, as shown in Fig. 2.31. When the interaction is sufficiently weak, two doublets ($|++\rangle \leftrightarrow |-+\rangle$; $|--\rangle \leftrightarrow |+-\rangle$ and $|++\rangle \leftrightarrow |+-\rangle$; $|--\rangle \leftrightarrow |-+\rangle$) are observed for these allowed transitions, and the distance between the resonant modes in each doublet is in the first approximation equal to the exchange energy J. As the exchange interaction increases, the external branches

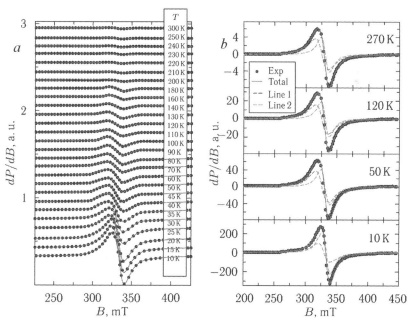

Fig. 2.30. Evolution of the ESR absorption spectra with a temperature variation for the $Cs_2Cu(VO)(P_2O_7)_2$ powder sample (*a*): ESR spectra decomposition into two absorption lines at various temperatures (*b*): the points are experimental data, the dashed lines are resolved resonance modes from approximations by the sum of two Lorentzians, solid lines the sum of two components.

(transitions $|++\rangle \leftrightarrow |-+\rangle$ and $|--\rangle \leftrightarrow |-+\rangle$) diverge and weaken in amplitude, while the internal branches (transitions $|--\rangle \leftrightarrow |+-\rangle$ and $|++\rangle \leftrightarrow |+-\rangle$) shift to the central frequency $h\nu = \frac{1}{2}\left(g_{jz} + g_{jz}\right)\mu_B B_z$ and their intensity increases substantially, so that for a certain value of the exchange parameter in the spectrum only two internal modes are observed. In this case, the values of the effective g-factors can deviate significantly from the values expected on the basis of the theory of the crystal field for a given ligand coordination.

The temperature dependences of the main parameters of the ESR spectra are shown in Fig. 2.32 for both resolved resonant components. A small deviation of the g-factor below ~ 50 K indicates an increase in the role of short-range order correlations and the appearance of the corresponding internal fields. The same behaviour of the individual components for both lines speaks for a strong exchange interaction between the copper and vanadium subsystems.

It should be noted that the line width ΔB exhibits a break at $T \sim 22$ K, which is most pronounced for the component ΔB_1. This result is also in good agreement with the data on the static magnetic susceptibility, which shows an anomaly at the same temperature.

The temperature dependences of the magnetic susceptibility in a field of 0.1 T are shown in Fig. 2.33. In the whole investigated temperature range, $\chi(T)$ exhibits Curie–Weiss growth and does not

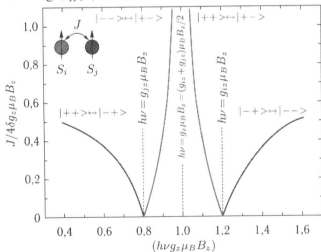

Fig. 2.31. Schematic diagram of the spectrum of 2 ions i and j (with different g-factors), connected by an isotropic exchange interaction J, described by the Hamiltonian (2.12), when the difference between their Zeeman energy is $h\nu = \frac{1}{2}\left(g_{jz} + g_{jz}\right)\mu_B B_z = 2\delta g_z \mu_B B_z \gg J$. The diagram is constructed for the case $\delta g_z / g_z = 0.2$.

exhibit long-range magnetic order as the temperature is lowered to 2 K. The best description of $\chi(T)$ is obtained with the parameters $\chi_0 = -4.07 \cdot 10^{-4}$ emu/mol, $C = 1.08$ emu K/mol and $\Theta = -14$ K. The experimental data plotted on the scale $(\chi-\chi_0)$ $(T-\Theta)$ (the inset in Figure 2.33) shows a weak anomaly at $T^* = 22$ K, which indicates the intersection of dashed lines. In addition, a significant deviation from the Curie–Weiss law at $T < 100$ K and a substantial increase of χ in the low-temperature region can be clearly seen on this scale.

The square of the effective magnetic moment, calculated from the Curie constant, is $\mu_{eff}^2 = 8.64$ μ_B^2 and agrees satisfactorily with the theoretical estimate of 9.13 μ_B^2 for effective g-factors $g_{Cu} = 2.037$ for Cu^{2+} ions $(S = 1/2)$ and $g_V = 2.001$ for V^{4+} $(S = 1/2)$. The negative value of the Weiss temperature indicates the dominance of antiferromagnetic exchange interactions in the paramagnetic region, however, the low-temperature growth of χ indicates the presence of the ferromagnetic contribution to the exchange correlations.

The Cu^{2+} ions $(d_{x^2-y^2})$ in a planar-square environment and V^{4+} (d_{xy}) in a square-pyramidal environment have magnetoactive orbitals in

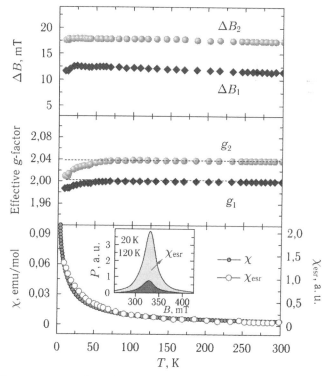

Fig. 2.32. The temperature dependences of the ESR line width ΔB (upper part), the effective g-factor (middle part), and the integral ESR intensity χ_{ESR} (lower part).

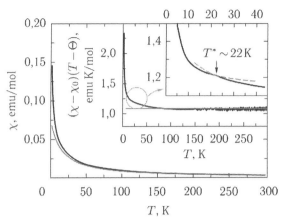

Fig. 2.33. Temperature dependences of the magnetic susceptibility of the compound $Cs_2Cu(VO)(P_2O_7)_2$ in the field $B = 0.1$ T. The solid curves are the result of the Curie–Weiss approximation. On the inset: the temperature dependence $(\chi-\chi_0)(T-\Theta)$.

the basal plane [95]. Thus, it can be expected that Cu–O–V bonds directed perpendicular to the basal plane do not play an important role in magnetic exchanges. At the same time, magnetoactive orbitals can participate in exchanges through phosphate groups [96]. The appearance of the ferromagnetic component in the exchange interactions can be related to closely located vanadyl pyramids for ions V1 and V3, which are located at a distance of 4.06 Å from each other. The possibility of the appearance of a ferromagnetic interaction between vanadyl pyramids at a distance of ≤ 4.5 Å from each other is shown in [97], and the intensity this interaction decreases with increasing distance between ions, and, ultimately, it is replaced by an antiferromagnetic one. The exchange interaction between other atoms of vanadium and copper may have an antiferromagnetic nature [98, 99]. The partial crystallo-chemical disorder that is present in the system under investigation because of the interchange of Cu1/V1, Cu2/V2 and Cu3/V3 weakens the anomaly at T^*, which can be interpreted as the ordering temperature. The presence of various competing exchanges and partial structural disorder can lead to a cluster type of magnetic ordering in the test compound at a temperature of about 22 K.

2.2.5. Bismuth–iron selenite–oxochloride

The crystalline structure of bismuth–iron selenite oxochloride $Bi_2Fe(SeO_3)_2OCl_3$ is a layered monoclinic, space group $P2_1/m$, as

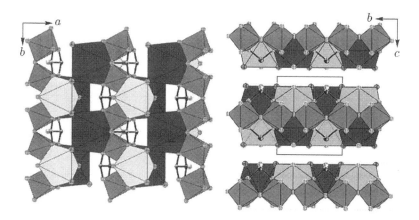

Fig. 2.34. The polyhedral appearance of the layered crystal structure of the compound $Bi_2Fe(SeO_3)_2OCl_3$.

shown in Fig. 2.34 [100]. The main feature of the crystal structure is the presence of isolated zigzag chains of Fe^{3+} ions ($S = 5/2$) along the b axis (Fig. 2.34) formed by corner bound octahedra FeO_6 surrounded by non-magnetic groups BiO_4Cl_3, BiO_3Cl_3 and SeO_3.

Evolution of the ESR absorption spectra for a powder sample of $Bi_2Fe(SeO_3)_2OCl_3$ with a temperature variation is shown in Fig. 2.35 [100]. A single exchange-narrowed line is observed in the spectra, apparently corresponding to a signal from Fe^{3+} ions in a low-level octahedral oxygen coordination. A satisfactory description of the shape of the ESR line is achieved when two resonance modes are taken into account, as shown in Fig. 2.35 a. Resonance modes differ significantly by intensity: the main contribution to the absorption is made by the much more intense mode L_1, whereas the contribution of the L_2 component is negligible ($< 2\%$) and is due to the presence of a small amount of impurity in the sample.

The main resonance absorption line of ESR is characterized by an isotropic effective g-factor of $g = 1.999$, typical for high-spin Fe^{3+} ions in the octahedral coordination. The quasi-one dimensional character of the behaviour of the magnetic subsystem is also visible in the temperature dependence of the static magnetic susceptibility in the field $B = 0.1$ T (Fig. 2.35 b), and in the behaviour of the integrated intensity of the electron paramagnetic resonance (Fig. 2.35 c). As the temperature is lowered, $\chi(T)$ passes through a wide maximum at $T \sim 130$ K, which corresponds to the formation of a regime of short-range field magnetic correlations. The line width ΔB_1 increases significantly with decreasing temperature, while its

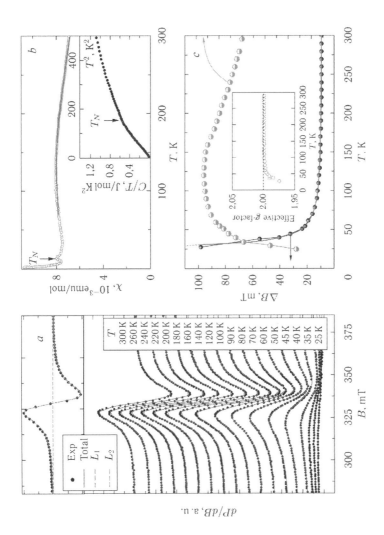

Fig. 2.35. Static and dynamic magnetic properties of $Bi_2Fe(SeO_3)_2OCl_3$; *a*) Evolution of ESR spectra with temperature variation: points – experimental data, lines – approximation by the sum of two Lorentzians. *b*) Temperature dependences of the magnetic susceptibility and specific heat (on the inset). *c*) Temperature dependences of line width, *g*-factor and integral intensity of ESR spectra.

amplitude passes through a maximum. As can be seen from Fig. 2.35 *b*, the temperature dependence of the magnetic susceptibility after passing through a wide correlation maximum demonstrates an anomaly corresponding to the establishment of a long-range antiferromagnetic order at $T_N = 13$ K. This anomaly is also seen in the data on the specific heat shown in the inset to Fig. 2.35 *b*, although it is rather weak, due to the fact that most of the entropy is released at high temperatures because of short-range order correlations.

In accordance with the peculiarities of the $Bi_2Fe(SeO_3)_2OCl_3$ structure, one can expect that the dominant interaction will be a superexchange in the chain of FeO_6 corner-shared octahedra. Interaction through apical oxygen is expected to be antiferromagnetic both for 90° and 180° of the $Fe^{3+} - O^{2-} - Fe^{3+}$ bond angle [54]. Thus, the antiferromagnetic chains of spins $S = 5/2$ with the exchange integral $J_\parallel$ determine the magnetic behaviour of $Bi_2Fe(SeO_3)_2OCl_3$ with a wide correlation maximum at $\chi(T)$ at high temperatures. The corresponding relation between the value of the magnetic susceptibility at the maximum point χ_{max}, the maximum temperature $T(\chi_{max})$, and the g-factor for the Heisenberg antiferromagnetic spin chain $S = 5/2$ is determined by the expression $\chi_{max} \cdot T(\chi_{max})/g^2 = 0.38$. In $Bi_2Fe(SeO_3)_2OCl_3$, however, this value is only 0.26. Such a low value indicates either the reduction of the spin magnetic moment of the iron atoms or the frustration of the dominant exchange interaction between the nearest neighbors inside the chain. In a simplified treatment, taking into account only one exchange parameter $J_\parallel$, which determines the properties of such a chain, the position of the maximum of the magnetic susceptibility χ_{max} is related to $J_\parallel$ as $k_B T(\chi_{max})/|J_\parallel| = 10.6$, which determines $J_\parallel = 12$ K. A more accurate estimate can be obtained within the framework of the Fisher model for an infinite chain of classical spins $S = \infty$ [101]. The result of the approximation of $\chi(T)$ at high temperatures by the sum of three terms, taking into account the temperature-independent contribution χ_0 and the contribution from C/T defects and impurities according to the formula

$$\chi = \chi_0 + \frac{C}{T} + \alpha\chi_{chain} \qquad (2.13)$$

is represented by the line in Fig. 2.35 *b*. The contribution of χ_0 was estimated by summing the Pascal constants for the atoms in $Bi_2Fe(SeO_3)_2OCl_3$ and amounting to $\chi_0 = -2.1 \cdot 10^{-4}$ emu/mol. The

Curie constant is $C = 4.7 \cdot 10^{-2}$ emu/mol K, which gives an estimate of the defect/impurity concentration ($S = 5/2$) at 1%. The coefficient $\alpha = 0.92$ indicates a deviation from the ideal 1D model. The value of the exchange parameter obtained from the approximation was $J_{\parallel} \approx 17$ K.

The interchain interaction $J_{\perp}$ can be estimated from relation

$$\exp\left(\frac{2|J_{\parallel}|}{k_B T_N}\right) = \frac{4 + Z\eta}{Z\eta}, \tag{2.14}$$

where $\eta = J_{\perp}/J_{\parallel}$ and $Z = 4$ is the number of the nearest neighbours. Thus, this relation gives an estimate of $\eta = 0.08$ and $J_{\perp} \approx 1.4$ K.

Analysis of the temperature dependence of $\Delta B_1(T)$ within the theory of critical broadening of the absorption line by the formula [102–106] gives:

$$\Delta B(T) = \Delta B^* + A \cdot \left[\frac{T_N^{\mathrm{ESR}}}{T - T_N^{\mathrm{ESR}}}\right]^{\beta}, \tag{2.15}$$

where the temperature-independent first term ΔB^* describes the high-temperature limit of the width of the exchange-narrowed line, the second term is responsible for the critical behaviour as it approaches from above to the ordering temperature T_N^{ESR}, β is the critical exponent. The use of formula (2.15) made it possible to describe satisfactorily the experimental data over the entire temperature interval (solid curve in Fig. 2.35 c). The best agreement was reached when following parameters of the model: $\Delta B^* = 10 \pm 1$ mT, $A = 42 \pm 1$ mT, $T_N^{\mathrm{ESR}} = 13 \pm 2$ K and $\beta = 1.75 \pm 0.5$. It can be seen that the value of the T_N^{ESR} parameter is in good agreement with the value of the Néel temperature, at which an anomaly in the magnetic susceptibility and specific heat is observed (Fig. 2.35 b), and the value of the critical exponent fully corresponds to the theoretically expected value of $\beta = 7/4$ for 1D AFM in the Kawasaki theory for the spin chain as the magnetic ordering temperature is approached.

First-principles calculations of the basic exchange interactions of Fe–Fe were carried out taking into account the hybridization of Fe-d orbitals with O-p, Cl-p, Se-p and Bi-p by constructing effective Wannier Fe-d orbitals and hopping integrals. As expected from structural considerations and experimental data, intra-chain interactions between the nearest neighbours $t_{\parallel}$, the interaction between the neighbors following the nearest $t'_{\parallel}$ and interchain

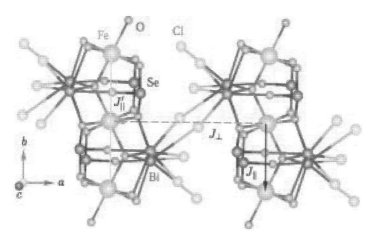

Fig. 2.36. The main mechanisms of exchange interactions $J_\parallel$, $J'_\parallel$ and $J_\perp$ in the compound $Bi_2Fe(SeO_3)_2OCl_3$.

exchange $t_\perp$ were the dominant exchange interactions. The main paths of exchange interactions are shown in Fig. 2.36. It can be seen that the exchange of $J_\parallel$ is carried out by supersuper-exchange path of Fe–O–Fe, where the cation-anion bond angle is 128.6°, whereas $J'_\parallel$ goes through a supersuper exchange mechanism involving Fe, O, Se and Bi ions. Interchain exchange interaction $J_\perp$ goes along the long path of Fe–O–Bi1–Cl–Bi2–O–Fe. All the calculated values of the exchange integrals turned out to be antiferromagnetic and amount to $J_\parallel \approx 21$ K, $J'_\parallel \approx 4.6$ K, and $J_\perp \approx 0.6$ K, which confirms the picture of the quasi-one-dimensional spin model for the compound studied. The ratio of the exchanges between nearest neighbors to the exchange between the next-nearest neighbors $\kappa = J'_\parallel / J_\parallel \approx 0.22$ turns out to be close to the critical value ($\kappa_{cr} = 0.25$) separating the collinear ($\kappa < \kappa_{cr}$) and the helicoidal ($\kappa > \kappa_{cr}$) magnetic structures [107], so we can expect that the magnetic structure of $Bi_2Fe(SeO_3)_2OCl_3$ will be collinear.

2.3. Alternating chain of half-integer spins

The alternating chain of half-integer spins $S = 1/2$, the scheme of which is shown in Fig. 2.37, can be described by three equivalent Hamiltonians [33]:

$$\hat{H} = \sum_i J_1 \vec{S}_{2i-1}\vec{S}_{2i} + J_2 \vec{S}_{2i}\vec{S}_{2i+1} = \sum_i J_1\vec{S}_{2i-1}\vec{S}_{2i} + \alpha J_1\vec{S}_{2i}\vec{S}_{2i+1} =$$
$$= \sum_i J(1+\delta)\vec{S}_{2i-1}\vec{S}_{2i} + J(1-\delta)\vec{S}_{2i}\vec{S}_{2i+1}, \qquad (2.16)$$

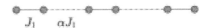

J_1 αJ_1

Fig. 2.37. Schematic model of a chain with alternating interaction.

where

$$J_1 = J(1+\delta) = \frac{2J}{1+\alpha}, \quad J = \frac{J_1 + J_2}{2} = J_1\frac{1+\alpha}{2},$$

$$\alpha = \frac{J_2}{J_1} = \frac{1-\delta}{1+\delta}, \quad \delta = \frac{J_1}{J} - 1 = \frac{J_1 - J_2}{2J} = \frac{1-\alpha}{1+\alpha},$$

with antiferromagnetic interactions $J_1 \geq J_2 \geq 0$ and the alternation parameters $0 \leq \alpha, \delta \leq 1$.

For the values of the alternation parameter $0 \leq \alpha \leq 0.9$, the value of the spin gap Δ in such a chain is determined as [33, 53]

$$\Delta^*(\alpha) = \frac{\Delta(\alpha)}{J_1} = (1-\alpha)^{3/4}(1+\alpha)^{1/4}, \quad \Delta^*(\delta) = \frac{\Delta(\delta)}{J} = 2\delta^{3/4}. \quad (2.17)$$

Analytic formulas for alternating chains were also obtained in the one-magnon approximation for describing the temperature dependences of specific heat and magnetic susceptibility as $T \to 0$ [33]:

$$\frac{\chi J_1}{N_A g^2 \mu_B^2} = \frac{1}{\sqrt{2\pi} f(\Delta^*)}\sqrt{\frac{\Delta}{k_B T}}e^{-\Delta/k_B T}, \quad (2.18)$$

$$\frac{C}{N_A k_B} = \frac{3}{\sqrt{2\pi}}\frac{\Delta/J_1}{f(\Delta^*)}\left(\frac{\Delta}{k_B T}\right)^{3/2}\left[1+\frac{k_B T}{\Delta}+\frac{3}{4}\left(\frac{k_B T}{\Delta}\right)^2\right]e^{-\Delta/k_B T}, \quad (2.19)$$

where the function $f(\Delta^*)$, shown in Fig. 2.38, can be replaced by a constant for a very small gap, when $\alpha \sim 1$ ($\delta \ll 1$), and in the interval $0 < \alpha < 0.4$ can be described by a polynomial:

$$f(\Delta^*) = \frac{\pi}{2} - 0.034289(\Delta^*) - 1.18953(\Delta^*)^2 + 0.40030(\Delta^*)^3. \quad (2.20)$$

Numerical calculations of the magnetic susceptibility and specific heat of alternating Heisenberg chains over a wide temperature range were performed in [12]. The results of these calculations are shown

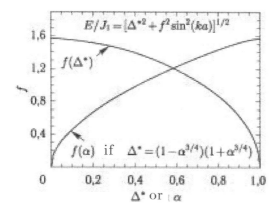

Fig. 2.38. The function $f(\Delta^*)$, obtained in the one-magnon approximation for $S = 1/2$ of the antiferromagnetic alternating chain, where the spin gap $\Delta^* = \Delta/J_1$ [33].

in Fig. 2.39. For any values of the parameter α, a broad maximum exists on the curves $\chi(T)$ and $C(T)$. The position of the maximum on the $\chi(T)$ dependence is practically independent of α, whereas on the $C(T)$ dependence the position of the maximum shifts to the right with increasing α.

For a wide range of temperatures, we can use the analytical dependence $\chi(T)$ in the Hamiltonian's notation via $2J$ as a polynomial:

$$\chi(T) = \frac{Ng^2\mu_B^2}{|J|} \frac{A_0\left(\frac{k_BT}{|J|}\right)^2 + A_1\frac{k_BT}{|J|} + A_2}{\left(\frac{k_BT}{|J|}\right)^3 + B_1\left(\frac{k_BT}{|J|}\right)^2 + B_2\frac{k_BT}{|J|} + B_3}, \qquad (2.21)$$

where the numerical values of the coefficients A_i and B_i in the ranges $0 \le \alpha \le 0{,}4$ and $0.4 < \alpha \le 1$ are given in [108].

The compound $(VO)_2P_2O_7$ contains two closely spaced chains of magnetically active ions V^{4+}. Previously, it was regarded as a spin ladder, but according to the data of later works it should rather be considered as a compound in which there are two types of alternating spin chains. This model is described below. The monoclinic crystal structure $(VO)_2P_2O_7$ (the space group $P2_1$) is shown in the left panel of Fig. 2.40. The elementary cell contains eight non-equivalent positions for vanadium and phosphorus. Alternate chains of two types **A** and **B** are directed along the b axis, each of them contains pairs of edge-shared pyramids VO_5, which are separated by PO_4 tetrahedra [109].

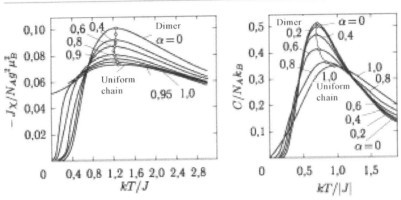

Fig. 2.39. Calculation of the magnetic susceptibility (left panel) and specific heat (right panel) for chains with a half-integer spin with varying degrees of alternation [12].

At the lowest temperatures, a peak corresponding to the nonmagnetic ground state of this compound was found in the spectra of ^{31}P nuclear magnetic resonance $(VO)_2P_2O_7$, as shown in the right panel of Fig. 2.40. With increasing temperature, this peak shifts and splits first into two, and then into four lines. That is, eight phosphorus positions are divided into four groups that sense the various internal fields created by vanadium ions. Then, at the highest temperatures, only one peak is observed again in the spectrum. The displacement and the non-monotonic change in the amplitudes of the individual lines with increasing temperature indicate the presence of several spin components with different temperature dependences of the magnetic susceptibility. The gap estimate for each of the four spin components was calculated from the low-temperature Knight shift for the quadratic dispersion law of magnons. Four lines in the spectrum of nuclear magnetic resonance are divided into two groups, differing in the temperature direction of the Knight shift and the size of the gap. For one group of peaks $\Delta = 35$ K, for another group $\Delta = 52$ K. In contrast, only one peak was observed in the ^{51}V spectrum, and the gap value, determined from its Knight shift, was $\Delta = 68$ K. The discrepancy in the values of the gaps determined from the Knight shift ^{31}P and ^{51}V, may be related to with the fact that the internal fields of vanadium ions belonging to different chains act on the phosphorus ions [109].

As shown in Fig. 2.41, in the experiments on inelastic neutron scattering [110], two dispersion curves and, correspondingly, two gaps $\Delta_1 = 36$ K and $\Delta_2 = 69$ K were observed. Both dispersion

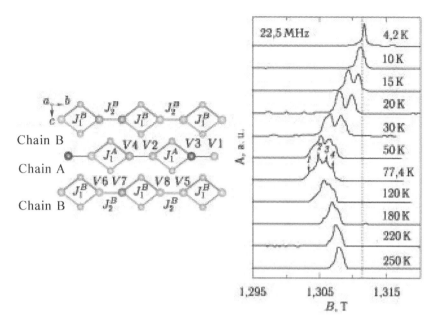

Fig. 2.40. Left panel: a schematic representation of the crystal structure $(VO)_2P_2O_7$. Right panel: the spectra of nuclear magnetic resonance ^{31}P in $(VO)_2P_2O_7$ with a temperature variation [109].

curves indicate a large value of the antiferromagnetic exchange parameter along the *b* axis. At the boundary of the Brillouin zone in this direction, the energies of the two branches are compared and reach $E \sim 180$ K. The structural unit cell coincides with the magnetic unit cell, so the dispersion of magnetic excitations is observed in the entire Brillouin zone. Using the dispersion law for

the alternating chain $E^2 = \Delta^2 + \left(\dfrac{\pi J}{2}\right)^2 \sin^2(q)$ (q is the wave vector) and the energy values of the two branches at the Brillouin zone boundary, we can estimate the mean antiferromagnetic exchange parameter $J = (J_1 + J_2)/2$ for of each of the two chains. The scale of alternation in each chain δ is determined from the expression $\Delta(\delta)/J \approx 2\delta^{3/4}$ [111].

The field dependence of the magnetization of $(VO)_2P_2O_7$ also indicates the presence of two gaps in the spectrum of magnetic excitations of this compound at low temperatures [111], as shown in the right panel of Fig. 2.41. The $M(B)$ curve exhibits bends at critical values of the magnetic field $B_{C1} = 25$ T and $B_{C2} = 46$ T. From the ratio $\Delta_{1,2} = g\mu_B B_{C1,2}/k_B$, it is possible to determine the gap values

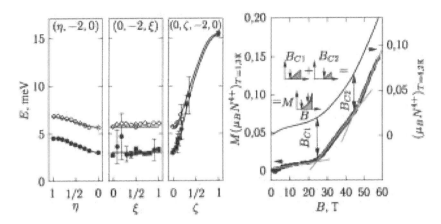

Fig. 2.41. Left panel: Dispersion curves of magnetic excitations in $(VO)_2P_2O_7$. The right panel: the field dependences of the magnetization in $(VO)_2P_2O_7$ [110, 111].

as $\Delta_1 = 33$ K and $\Delta_2 = 62$ K, which agrees well with the data on inelastic neutron scattering.

2.4. A homogeneous chain of integer spins

A homogeneous antiferromagnetic chain $S = 1$ with single-ion anisotropy does not undergo long-range magnetic order, but has a gap in the spectrum of magnetic excitations. The Hamiltonian of such a chain can be written as [112, 113]

$$\hat{H} = |J| \sum_i \left(\hat{S}_i^z \hat{S}_{i+1}^z + \lambda \hat{S}_i^z \hat{S}_{i+1}^z + \mu \left(\hat{S}_i^z \right)^2 \right), \qquad (2.22)$$

where λ and μ are the coefficients of anisotropy and splitting of the levels of an individual ion in a zero field.

The description of low-energy excitations by magnons alone is incorrect for systems with easy-axis anisotropy, and it is necessary to take into account the existence of solitons. If the magnon in the case of strong anisotropy was interpreted as the overturning of one spin, then the soliton is a collective excitation of the entire system. Spin reversal occurs not on one site, but includes some (always odd) number of spins. Since the soliton contains an odd number of lattice spins, the permissible spin of the soliton depends on the chain in which it is formed. The spin of a soliton will be integer if the spins in the chain are integer, and half-integer if the spins in the chain are half-integer [113].

The estimate of the gap in the chain $S = 1$ is made in a number of theoretical works and is determined by the expression $\Delta = 0.41J$ [114]. For $\mu = 0$, the gap exists in the interval $0 \leq \lambda \leq 1.18$. For $\mu \neq 0$, the gap differs from zero only for $-0.25 \leq \mu/2J \leq 0.8$. For $\mu < 0$, the gap rapidly decreases with increasing absolute value μ [115].

To describe the temperature dependence of the magnetic susceptibility, the following formula can be used:

$$
= \frac{N_A g^2 \ _B^2 S(S\ 1)}{k_B T} \exp\left(-\frac{}{k_B T}\right) \frac{\sum \left(\frac{}{k\ T}\right)}{\sum \left(\frac{}{k\ T}\right)}, \tag{2.23}
$$

where Δ is the gap between the singlet and triplet states, J is the antiferromagnetic interaction in the chain, and the numerical coefficients A_i and B_i are given in [116]. The form of the $\chi(T)$ dependence in dimensionless coordinates is shown in Fig. 2.42 (left).

The temperature dependence of specific heat of the antiferromagnetic chain $S = 1$ was calculated by the Monte Carlo method [116]. The $C(T)$ curve exhibits a broad maximum with a temperature $T_{max} \sim 2\Delta$ (0.8J). An example of a theoretically calculated specific heat on chains of different lengths is shown in the right panel of Fig. 2.42.

The PbNi$_2$V$_2$O$_8$ system is the realization of the Haldane disordered antiferromagnetic chain $S = 1$. The crystal structure and magnetic susceptibility of the compound PbNi$_2$V$_2$O$_8$ are shown in Fig. 2.43. This material has a tetragonal structure, the space group of symmetry

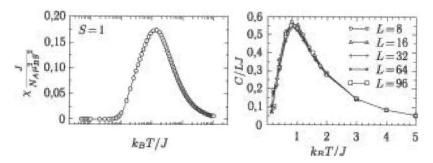

Fig. 2.42. The theoretically calculated temperature dependences of the magnetic susceptibility (the left panel) and the specific heat(right panel) for antiferromagnetic chains $S = 1$ [116, 117].

$I4_1/acd$, in which the edge-shared octahedra NiO_6 form helical chains along the c axis. These chains are separated by ions Pb^{2+} and tetrahedrons VO_4. The distance between the nearest Ni ions in the chain is 2.8 Å, between the chains 05.9 Å [118]. The temperature dependence of the magnetic susceptibility $PbNi_2V_2O_8$ exhibits a broad maximum at $T \sim 120$ K below which the susceptibility decreases exponentially as $T \rightarrow 0$ K. The evaluation of the exchange interaction in the chain from the high-temperature region is $J = 95$ K, which determines the gap in the spectrum of magnetic excitations $\Delta \approx 39$ K. This value is in good agreement with the data of neutron diffraction studies ($\Delta = 46$ K) and an estimate of the magnetic susceptibility decrease ($\Delta \sim 30$ K). Allowance for the interaction between the chains $J' \sim 1$ K and the anisotropy parameter $D = 3$ K makes it possible to obtain better agreement between the calculations and the results of the experiment. According to these calculations, the ground state of the $PbNi_2V_2O_8$ system on the phase diagram J'–D shown in Fig. 2.44, is in the spin-gap disordered phase, but close to the boundary of the spin-gap and ordered phases [119].

In the isostructural compounds $PbNi_2V_2O_8$ and $SrNi_2V_2O_8$, strong gap suppression was observed in comparison with the expected value: $\Delta = 0.23\,J$ and $0.25\,J$, respectively. That is, by the ratio of the parameters, both compounds are located near the boundary of the 'spin liquid–antiferromagnet' phases on the phase diagram in Fig. 2.44 [120] which is also called the Sakai–Takahashi diagram. The influence of the anisotropy D and the interaction between the

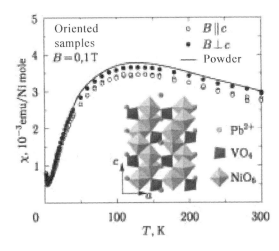

Fig. 2.43. Temperature dependences of the magnetic susceptibility of the compound $PbNi_2V_2O_8$. The inset shows a fragment of its crystal structure [118].

Haldane chains J' are taken into account in the Hamiltonian of the following form:

$$\hat{H} = J\sum_{\langle i,j\rangle} S_i S_j + D\sum_j \left(S_j^z\right)^2 + J'\sum_{(i,j)} S_i S_j. \qquad (2.24)$$

Single crystals of $SrNi_2V_2O_8$ were used to study the magnetic susceptibility, specific heat, and experiments on inelastic neutron scattering which showed that in this compound (similar to the Pb-based compound) the ground state is a spin singlet with a gap [121]. The parameters of this compound are $J = 8.9$ meV, $D = -0.51$ meV.

An analysis of the behaviour of quasi-one dimensional systems with integer spin can be carried out using the approaches developed to describe the Bose–Einstein condensation of magnons. In systems with an energy gap, that is, in systems with a non-magnetic ground state, the application of the magnetic field splits the upper level (triplet with $S = 1$) into three sublevels +1, 0 and −1 in accordance with the Zeeman effect. When the critical field B_C is reached, the lower sublevel (−1) becomes the ground state. That is, in a magnetic field the ground state is not a spin singlet but a state with a magnetization ($S = -1$) and with gapless magnons. This transformation is a quantum phase transition induced by an external magnetic field. The situation with the Bose–Einstein condensation of magnons was noted in $BaCuSi_2O_6$, $NiCl_2$–$4SC(NH_2)_2$, $TlCuCl_3$ [23]. Also, under the action of a magnetic field, the formation of a superlattice of localized triplets ($S = 1$) on a grid of orthogonal dimers in $SrCu_2(BO_3)_2$ was observed. The magnetic field acts as

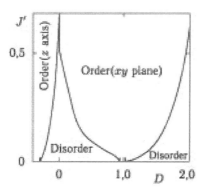

Fig. 2.44. The phase diagram for the quantum ground state of a chain with integer spin, taking into account the anisotropy D and the interaction between Haldane chains J' (Sakai–Takahashi diagram) [120].

a chemical potential, and magnons are excited only at $B > B_C$. If the magnons do not completely interact with each other, then the excitations can be collected in a condensate from free bosons (the phenomenon of Bose–Einstein condensation in its pure form), which corresponds to antiferromagnetic XY ordering. If the interaction between the magnons is very strong, then we can imagine a nucleus of bosons and a gas of free excitations (fermions) with a Fermi surface. In reality, the interaction between the magnons will be somewhere in the middle between the two extreme cases, and most systems with an energy gap can be considered as an interacting 'sea' of fermions. Moreover, some sources state that even a weak anisotropy can qualitatively change the behaviour of the system in comparison with what is expected from Bose–Einstein condensation.

In the Haldane chains, it is initially assumed that the magnons interact, which means a fundamental difference of the ground state in the fields $B > B_C$: instead of the magnon condensate (bosons), the Luttinger gapless spin liquid is realized. In addition, the presence of magnetic anisotropy and interactions between chains can significantly change the nature of the field-induced phase transition in Haldane systems, which complicates the behaviour of physical properties and the phase diagram.

From the objects described in the literature with planar anisotropy $(D > 0)$ in the compound $Ni(C_2H_8N_2)_2NO_2ClO_4$ (NENP), the transition induced by the magnetic field was not observed up to temperatures of 0.2 K and fields of 13 T. On the contrary, the spin gap was retained even at B_C (the B_C value is determined from the appearance of nonzero magnetization, at $T = 1.3$ K $B_C = 13$, 7, 5, and 11 T in the axes a, b, and c, respectively). This result is due to the presence of an alternating field at the Ni^{2+} positions, since NENP has two non-equivalent positions for this ion. This staggered field leads to the presence of a small energy gap near the critical field B_C, resulting in a smooth crossover from the spin-gap singlet state to the state with finite magnetization instead of the phase transition.

The magnetic field-induced phase transition to a state with a magnetic order is observed in $Ni(C_5H_{14}N_2)_2N_3(ClO_4)$ (NDMAZ) and $Ni(C_5H_{14}N_2)_2N_3(PF_6)$ (NDMAP) containing Haldane chains of Ni^{2+} ions and also characterized by planar anisotropy $(D > 0)$. In experiments on neutron scattering and electron paramagnetic resonance, unusual spin excitations were observed, namely, in the ordered state, three different types of excitations were found. In the

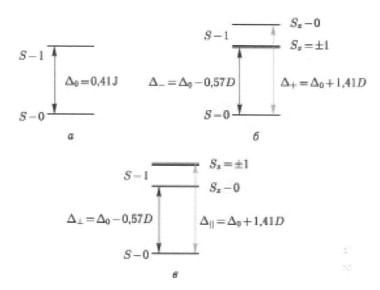

Fig. 2.45. Influence of anisotropy on the energy gap in the Haldane magnetic material. *a)* a gap in the Haldane system without anisotropy. *b)* two slits in the Haldane system, taking into account the planar anisotropy $D > 0$. *c)* two slits in the Haldane system with uniaxial anisotropy $D < 0$.

case of Néel ordering, the dominant excitations are the spin-wave modes.

The effect of anisotropy on the energy gap in the Haldane magnet is shown schematically in Fig. 2.45: the splitting of the triplet state excited by a triply degenerate state with respect to $S_z = -1, 0, 1$ leads to the appearance of two gaps: the first between the ground state $|0,0\rangle$ and the excited state $|1,\pm1\rangle$, the second between the ground state $|0,0$ and excited by $|1,0\rangle$. Depending on the nature of the anisotropy, the triplet levels with $S_z = 0$ and $S_z = \pm1$ will change places. For a planar anisotropy $D > 0$ (Fig. 2.45 *b*), a doubly degenerate state with a projection $S_z = \pm1$ will shift downward in energy, up, the state with $S_z = 0$ is lifted up, with the gaps being equal, small gap $\Delta_- = \Delta_0 - 0.57D$, between $|0,0\rangle$ and $|1,\pm1\rangle$ and a large gap $\Delta_+ = \Delta_0 + 141D$. For a uniaxial anisotropy $D < 0$ (Fig. 2.45 *c*), the notation of the gaps must be reversed ($\Delta_\parallel \to \Delta_+$, $\Delta\perp \to \Delta_-$).

In the $PbNi_2V_2O_8$ compound with uniaxial anisotropy ($D < 0$), the ordering induced by the external magnetic field occurs, as shown by experiments on magnetization in oriented powder samples [122]. At $T = 2.4$ K, two critical fields $B_{C\perp} = 14$ T and $B_{C\parallel} = 19$ T were recorded. The behaviour of the magnetization at $B > B_C$ was explained within the framework of the concept of the Bose–Einstein condensation of magnons, and it was asserted that the magnon picture of Bose–Einstein condensation is suitable for all Haldane systems without exception. Experimentally, a minimum is observed on the $M(T)$ curve in fields $B > B_c$ with a sharp inflection at a temperature T_{min}. As the temperature is lowered, the $M(T)$ dependence has a convex shape, as shown in the left panel of Fig. 2.46 b. T_{min} is considered the temperature of three-dimensional magnetic ordering. The increase in the magnetization at $T < T_{min}$ is due to the increase in the number of condensed magnons upon cooling. However, according to electron paramagnetic resonance, the temperature dependence of the spectra of triplet excitations was observed, which is explained by the interactions between them. Diffraction of neutrons on a powder also indicates a significant interaction between the chains and anisotropy.

In the context of magnetic field-induced phase transitions in systems with anisotropy and interchain interactions, the $SrNi_2V_2O_8$ compound is of considerable interest. First, it has a uniaxial anisotropy of such a scale that the splitting of the triplet state by the field is comparable to the energy of the Haldane gap; secondly, there is interaction between the chains, and thirdly, there is an ambiguous interpretation of the ground state: ordered or spin-liquid. It was established that the ground state of the previously considered compound $PbNi_2V_2O_8$ is a spin liquid [118], while the parameters D and J' the spin liquid-antiferromagnet on the Sakai–Takahashi diagram is very close to the phase boundary. In $SrNi_2V_2O_8$ at $T_N = 7$ K, a long-range order with weak ferromagnetism is formed, and the parameters D and J' are close to the phase boundary from the side of the antiferromagnetic order. The last assertion was refuted in [123]: according to nuclear magnetic resonance data on the $SrNi_2V_2O_8$ powder, the ground state is a spin singlet, and the transition at 7 K was not detected either in specific heat or in the magnetic susceptibility of this compound.

Studies of static magnetic susceptibility, neutron diffraction, field dependences of magnetization and specific heat in the region of ultralow temperatures on single crystals of $SrNi_2V_2O_8$ [121] confirm that the ground state is a spin singlet with a gap between it and a triplet excited state. The splitting of the triplet state due to anisotropy was observed in neutron diffraction, the ratio of the mode intensities indicates that the easy axis is directed along the crystallographic axis c. The phase transitions induced by the magnetic field occur at

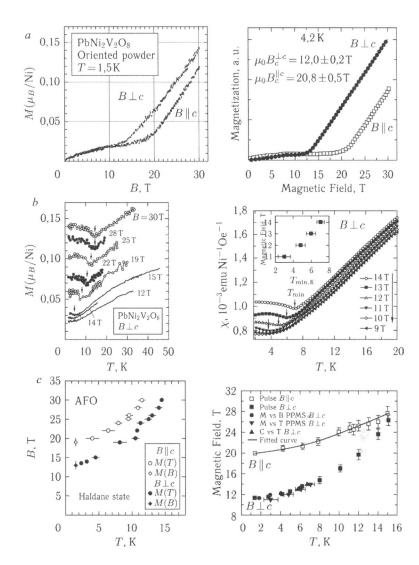

Fig. 2.46. Bose–Einstein condensation of magnons in $PbNi_2V_2O_8$ (left panels) [122] and in $SrNi_2V_2O_8$ (right panels) [123]. From the top down: *a*) the field dependence of the magnetization, *b*) the temperature dependence of the magnetization in the fields, *c*) the phase diagram in magnetic field–temperature coordinates.

$B_{C\perp c} = 12.0$ T and $B_{C\|c} = 20.8$ T at $T = 4.2$ K due to the fact that the Zeeman splitting closes the energy gaps. Two critical fields indicate an anisotropic Zeeman splitting of the triplet state. Two energy gaps between the ground and triplet states were also determined from the $M(B)$ dependences for two field directions. This confirms the idea that the anisotropy of an isolated ion partially removes the degeneracy of the triplet state.

Estimates from the neutron diffraction in $SrNi_2V_2O_8$ make the gap value $\Delta_\| = 1.57$ meV, $\Delta_\perp = 2.58$ meV, with $\Delta_0 = 2.29$ meV and $D = -0.51$ meV. The magnitude of the exchange integral in the chain is $J = 8.9$ meV. An estimate of the formula $\Delta = 0.41J$ leads to a value of this gap 3.65 meV, which exceeds the mean gap Δ_0. This difference is due to interchain exchange (according to the estimate [118] $J' \sim 0.18$ meV). The kinks in the dependences of $M(B)$ in $SrNi_2V_2O_8$ are observed at two values of the field: 12 T ($\perp c$) and 20.8 T ($\| c$), they correspond to the gap closing $\Delta(B_c) = 0$. The critical field along the direction of the c axis is greater than in the direction perpendicular to the c axis, i.e., an uniaxial anisotropy with an easy axis along c is confirmed. Using the magnitude of the critical field, the energy of the gap can be calculated in several ways. Two methods were used in [121]: the perturbation method (non-interacting bosons or fermionic approach), which gave estimates for the gap values $\Delta_\| = 0.9$ meV, and $\Delta_\perp = 2.69$ meV, and the boson model (interacting bosons), an estimate in which $\Delta_\| = 1.56$ meV, $\Delta_\perp = 2.69$ meV. It can be seen that the second model is much closer to the estimates from neutron diffraction.

The behaviour of the magnetization temperature dependences of $SrNi_2V_2O_8$ on the temperature in fields $B > B_c$ is similar to the behaviour of the magnetization in $PbNi_2V_2O_8$ described above, as shown in the right panel of Fig. 2.46 b [122]. This behaviour for $PbNi_2V_2O_8$ was described in the language of Bose–Einstein condensation of non-interacting magnons. However, in $SrNi_2V_2O_8$, appreciable anisotropy can lead to new physical processes. To understand the nature of the field-induced state in $SrNi_2V_2O_8$, specific heat of this compound in fields up to 14 T was investigated [121]. At $B = 0$, there are no anomalies in specific heat. In fields in the temperature region below 7 K, a fuzzy anomaly appears. At $B \perp c = 14$ T, a sharp peak at $T_N = 6.75$ K, corresponding to the transition to a state with a long-range magnetic order, is observed on the $C(T)$ curve.

From the position of specific heat and magnetization anomalies, a phase diagram was constructed in the coordinates $B–T$ for the two directions of the magnetic field, shown in the right panel of Fig. 2.46 c. The phase boundary is described by a power function in the form $T_N = A(B–B_c)^\phi$, where A is a constant that is the same for the cases $B \| c$ and $B \perp c$, and the exponent ϕ is different for these two directions [123]. For $B \parallel c$, the value of $\phi = 0.57$ turned out to be close to the value of 2/3 typical for the mechanism of the three-dimensional Bose–Einstein condensation of magnons. For $B \perp c$, $\phi = 0.43$, which is significantly smaller than the exponent of Bose-Einstein condensation. In the ideal case, spontaneous symmetry breaking of $SO(2)$ occurs in the case of the Bose–Einstein condensation of magnons, and the condensation phenomenon can occur only in tetragonal geometry with a field parallel to a single axis (in the case of $SrNi_2V_2O_8$ $B \parallel c$). For all other directions of the external field the transition belongs to the Ising universality class ($\phi = 1/2$). The resulting value of $\phi = 0.43$ for $B \perp c$ is close to 1/2.

The field-induced phase transition to a state with a three-dimensional long-range magnetic order contradicts the expected crossover between the states of a spin liquid with a gap and a spin liquid without a gap (Tomonaga–Luttinger liquid). This crossover was predicted for isolated Haldane chains in [124]. Unlike the Bose-Einstein condensation, the state of the Tomonaga–Luttinger liquid is disordered due to large quantum and thermal fluctuations in a one-dimensional magnet. The authors of [121] believe that the three-dimensional order in the external field is due to interchain interactions, which are not strong enough to form a long-range magnetic order without a field.

2.5. Spin-Peierls transition

A thoroughly studied metal oxide, which undergoes a spin-Peierls transition, is $CuGeO_3$. This transition is realized in crystals containing isolated homogeneous chains of half-integer spins. Such chains do not have a gap in the spectrum of magnetic excitations. However, the total energy of the crystal can be lowered by alternating the exchange interaction. At the spin-Peierls transition, the period of the crystal lattice doubles, which corresponds to a loss in the energy of the elastic subsystem, but a gap opens in the spectrum of magnetic excitations, which is equivalent to a gain in the energy of the magnetic subsystem.

The $CuGeO_3$ compound has an orthorhombic structure, the space symmetry group is *Pbmm* [125]. The chains of CuO_6 octahedra and the chains of GeO_4 tetrahedra that separate them are located along the *c* axis. The carriers of the magnetic moment $S = 1/2$ are Cu^{2+} ions with an unfilled *d*-shell.

The spin-Peierls transition was observed in $CuGeO_3$ in measurements of the magnetic susceptibility on single crystals [126]. As shown in Fig. 2.47, with a decrease in temperature at $T_M \sim 56$ K, a correlation maximum was observed, and at $T_{SP} \sim 14$ K a sharp decrease in the magnetic susceptibility was observed along all the crystallographic directions. The treatment of the $\chi(T)$ dependence at high temperatures in the model of a homogeneous chain of half-integer spins $S = 1/2$ [12] makes it possible to estimate the integral of the exchange interaction along the chain as $J_C \sim 88$ K from the relations (2.2). An estimate of the exchange interaction integrals with the next-nearest neighbors in the J_C' chain, as well as the interaction between the chains along the main crystallographic directions J_A and J_B, was carried out in neutron scattering experiments. For $J_C = 120$ K, the ratio J_C'/J_C was 0.2, $J_A = 0.1 J_C$, $J_B = 0.01\ J_C$ [127]. The last relationships characterize the one-dimensionality of the magnetic structure of $CuGeO_3$.

An estimate of the energy gap opening in the spectrum of magnetic excitations was carried out from the treatment of the $\chi(T)$ dependence at $T < T_C$ by the Bulaevsky model $\Delta(T) = 1.64\delta(T)\ J_C$, $\Delta(0) = 24$ K. In this case, the parameter of the alternation of the exchange integral in the chain δ was 0.17, and thus $J_C^1 / J_C^2 = 1.41$ ($J_C^{1,2}(T) = J_C\{1 \pm \delta(T)\}$). The spin-Peierls transition to $CuGeO_3$ is accompanied by the λ-type anomaly of in the specific heat[128]. Coefficients of thermal expansion in this compound exhibit anisotropic behaviour [129, 130]. Since the spin-Peierls transition is associated with static distortions of the crystal lattice, the period of the unit cell is doubled in X-ray diffraction experiments and neutron diffraction [131]. The largest displacements are experienced by Cu^{2+} ions, which move along the c

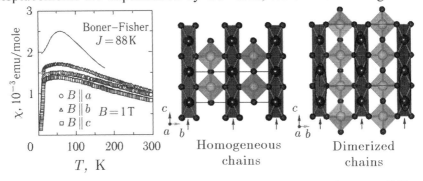

Fig. 2.47. Left panel: the temperature dependence of the magnetic susceptibility of $CuGeO_3$, the solid line is an approximation based on the Bonner–Fisher model. The middle and right panels: the scheme of atomic displacements during the spin-Peierls transition to $CuGeO_3$ [127].

axis, and O^{2-} ions in the *ab* plane. The values of the displacements that cause such radical changes in the magnetic properties of copper germanate are very small: $u^c_{Cu} \sim -0.0014$, $u^a_O \sim 0.0010$, $u^b_O \sim 0.0013$.

The resulting deformation of the lattice can be represented as an alternating rotation of GeO_4 tetrahedra around the axis that connects the apical oxygen ions. Such rotations cause alternating negative and positive displacements of copper ions along the *c* axis and, ultimately, small displacements of germanium ions along the *b* axis. Chains from one unit cell are dimerized in antiphase, as shown in the right panel of Fig. 2.47, the doubling of the period of the crystal lattice occurs not only along the *c* axis, but also along the *a* axis.

The magnetic phase diagram of $CuGeO_3$ is shown in the left panel of Fig. 2.48 [132]. Here the regions U, D and I correspond to a homogeneous, dimerized and incommensurate phase. The phase U is characterized by the presence of only short-range correlations in the chains. Phase D does not contain a macroscopic magnetic moment, which appears in phase I, where the periods of the magnetic and crystal structures are incommensurate. The incommensurate phase I differs in a period not only from the homogeneous phase U, but also from the dimerized phase D. Without a magnetic field at T_{SP}, a second-order phase transition occurs from phase U to D phase, and in the first critical field at low temperatures a first-order phase transition occurs from phase D to phase I. The critical field D–I $B = 12.5$ T ($T = 0$) corresponds to the transition temperature U–D at $T_{SP} = 14.3$ K ($B = 0$). On the magnetic phase diagram of $CuGeO_3$, the triple point is at 11.5 K and 13 T. When the spin-Peierls magnet

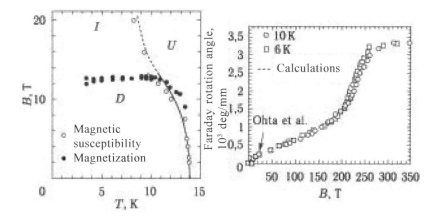

Fig. 2.48. The left panel: the magnetic phase diagram of $CuGeO_3$ [132]. The right panel: the field dependence of the Faraday rotation angle in $CuGeO_3$ [134].

is cooled in a field above the critical one, it undergoes a second-order phase transition from a homogeneous to an incommensurate magnetic phase.

The first critical field B_{C1} is proportional to the temperature of the spin-Peierls transition T_{SP} and the value of the gap in the spectrum of magnetic excitations Δ. The relationship between these parameters can be written as:

$$B_{C1} \sim 1.48 \frac{k_B T_{SP}}{g \mu_B} \sim 0.84 \frac{\Delta}{g \mu_B}. \tag{2.25}$$

In weak fields, the decrease in the critical temperature ΔT is proportional to the square of the magnetic field:

$$\frac{\Delta T}{T_{SP}} \sim -\left(\frac{g \mu_B B}{2 k_B T_{SP}(0)} \right)^2. \tag{2.26}$$

The values of the critical fields in different crystallographic directions differ due to the anisotropy of the g-factor ($g_a = 2.06$, $g_b = 2.27$, $g_c = 2.15$) [133]. At low temperatures, a hysteresis with a magnitude of about 0.1 T is observed in the field dependences of the magnetization of $CuGeO_3$ in the region of the first critical field. After the kink in the magnetization in the first critical field B_{C1} there is its monotonous growth, as shown in the right panel of Fig. 2.48. In a field of ~ 250 T, the magnetization of $CuGeO_3$ reaches saturation [134].

The picture of physical processes in spin-Peierls magnets when an external magnetic field is applied is as follows. In the first critical field B_{C1}, part of the dimers is destroyed (in $CuGeO_3$ of the order of 2%), and the resulting 'normal' magnetic moments are equidistantly distributed along the chain. The hysteresis of the $M(B)$ dependences in the region of the first critical field indicates, in particular, the possibility of pinning normal magnetic moments on structural defects. The number and distance between unpaired magnetic moments is determined by the magnetic field. As the magnetic field increases, the number of such moments increases, and the distance between them decreases.

The dimerized state is completely destroyed only in the second critical field B_{C2}, whose value is comparable to the doubled value of the exchange integral in the chain $2J_c$. For $B > B_{C2}$, the substance again appears in a homogeneous state, all magnetic moments in which are oriented in one direction.

Self-organization of an incommensurate phase of spin-Peierls magnets was observed in X-ray diffraction studies on the appearance of an additional period of a crystal lattice controlled by a magnetic field [135]. In an incommensurate phase, spatial modulation of the crystal lattice is established in the form of solitons with large dimensions: the half-width of the soliton is ~13.6 of the lattice period. Investigations of magnetostriction and thermal expansion coefficients in magnetic fields have shown that the lattice of solitons changes to sinusoidal modulation when approaching the I–U phase boundary [136].

2.6. Orbital mechanism of dimerization of the spin chain

Uniform antiferromagnetic chains $S = 1/2$, which have a noticeable entropy term in the energy, can reduce it due to dimerization over the orbital ordering mechanism. Such a scenario of formation of the ground state was detected in pyroxene $NaTiSi_2O_6$ (and also in $LiTiSi_2O_6$). This system has a monoclinic crystal lattice with a space symmetry group $C2/c$ [137]. In the structure of pyroxene there are isolated zigzag chains of edge-shared TiO_6 octahedra, as shown in Fig. 2.49. The chains are interconnected through the vertices of SiO_4 tetrahedra. At room temperature, only one titanium position is present in the $NaTiSi_2O_6$ structure. Thus, along the c axis, homogeneous antiferromagnetic chains of half-integer spins are formed.

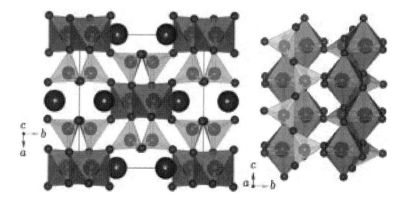

Fig. 2.49. Left panel: the crystalline structure of $NaTiSi_2O_6$. The octahedra are the structural units of $Ti^{3+}O_6$, SiO_4 tetrahedra, and individual spheres denote Na ions. Right panel: one-dimensional chain of TiO_6 octahedra twisted along the c axis [137].

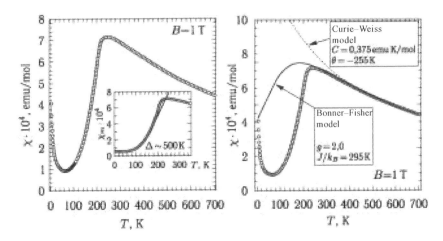

Fig. 2.50. Magnetic susceptibility of $NaTiSi_2O_6$. Left panel: symbols – experimental data, dotted line – approximate impurity contribution. On the inset: the susceptibility of the $NaTiSi_2O_6$ matrix χ_m, the solid line is the model calculation for $\Delta \sim 500$ K. Light panel: the dashed line is the high-temperature approximation according to the Curie–Weiss law, the solid line is the Bonner–Fisher model [80].

The temperature dependence of the magnetic susceptibility of $NaTiSi_2O_6$, measured in the field $B = 1$ T, is shown in Fig. 2.50 [80]. In the high-temperature region, the $\chi(T)$ dependence obeys the Curie–Weiss law; when the temperature is lowered, it exhibits a wide maximum and a sharp decrease at $T < 210$ K. Some growth of $\chi(T)$ at the lowest temperatures is due to impurities and defects. The inset to the left panel of Fig. 2.50 shows the susceptibility of the matrix $NaTiSi_2O_6$ χ_m, obtained by subtracting the impurity contribution χ_i. The value of χ_m sharply decreases practically to zero below 210 K and does not change at low temperatures. The residual susceptibility $\chi_0 = 6 \cdot 10^{-5}$ emu/mol is comparable in magnitude to the van Vleck contribution, observed earlier in other spin-gap systems in the ground state (NaV_2O_5, MgV_2O_5, CaV_2O_5 and CsV_2O_5). Thus, the ground state in $NaTiSi_2O_6$ can be described by a spin singlet with a gap $\Delta \sim 500$ K. This estimate of the gap size was made from the approximation of the low-temperature part of the magnetic susceptibility by the formula $\chi_m \sim \exp(-\Delta/k_B T)$.

At high temperatures, the $\chi(T)$ dependence obeys the Curie–Weiss law with the parameters: $C = 0.375$ emu K/mol, $\Theta = -255$ K. Below 400 K, the curve $\chi(T)$ deviates from this law toward smaller values, as shown in right panel in Fig. 2.50. The solid line was obtained from the Bonner–Fisher model for the antiferromagnetic chain of Heisenberg spins $S = 1/2$ with parameters $g = 2$, $J = 295$

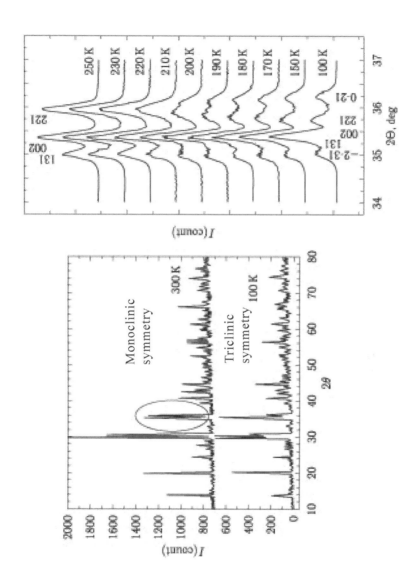

Fig. 2.51. The left panel: X-ray diffraction spectra of NaTiSi$_2$O$_6$, obtained at 300 K and 100 K. The right panel: X-ray reflections (−1 3 1), (0 0 2) and (2 2 1) with a temperature variation [80].

K, it well describes the experimental data above 250 K. However, at $T < 210$ K, the experimental dependence of $\chi(T)$ decreases more rapidly than follows from this model, which is due to the formation of the spin gap.

The temperature evolution of the X-ray spectra of $NaTiSi_2O_6$ with temperature is shown in Fig. 2.51. Some diffraction peaks, as shown in the right panel on an enlarged scale, split at $T < 210$ K, which indicates a decrease in crystal symmetry [80]. The crystal lattice parameters at $T = 100$ K in triclinic syngony were determined as: $a = 6.63$ Å,

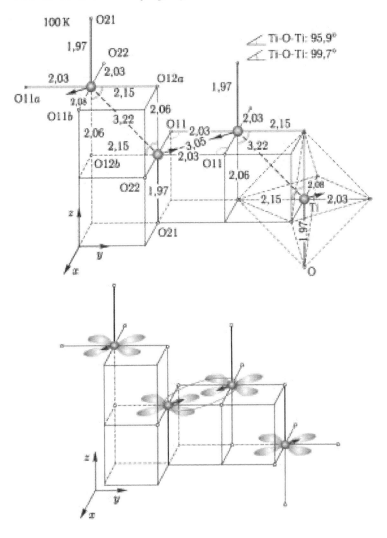

Fig. 2.52. The upper panel: alternating the chain of octahedra TiO_6 in $NaTiSi_2O_6$ at $T < 210$ K. The lower panel: the formation of dimers on the orbitals d_{xy}.

$b = 8.83$ Å, $c = 5.29$ Å, $\alpha = 90.2°$, $\beta = 102.3°$, $\gamma = 47.1°$. At room temperature this compound has a monoclinic crystal lattice with parameters: $a = 9.692$ Å, $b = 8.874$ Å, $c = 5.301$ Å, $\beta = 106.85°$, $\gamma = 47.1°$.

Despite some indications of the spin-Peierls transition to $NaTiSi_2O_6$, the structural phase transition here is associated with orbital ordering. In the structure of this compound, the magnetically active titanium ions Ti^{3+} in a distorted octahedral environment have two degenerate t_{2g} orbitals, on which one electron is located, which provides a magnetic interaction along the chain. In this case, the exchange magnetic interaction occurs with the participation of both orbitals. Such a Jahn–Teller situation is unstable, and below the $T_S = 210$ K the system undergoes a structural phase transition that removes the degeneracy of the t_{2g} orbitals in such a way that at low temperatures the electron occupies the lowest-lying orbital d_{xy}. This transition leads to a strong alternation of the exchange interactions in the chain, as shown in Fig. 2.52. In the case of overlapping orbitals d_{xy}, dimers appear in the chain. The transition to the singlet state leads to a sharp decrease in the magnetic susceptibility at T_S.

Spin ladders

3.1. Spin ladders with an odd number of legs: a spin liquid without a gap in the spectrum of spin excitations

A special class of low-dimensional magnets are systems of interacting antiferromagnetic chains, also called spin ladders. Thus, in Ref. [138], in calculations by the renormalization group method, it was shown that in the antiferromagnetic ladder $S = 1/2$, composed of an odd number of legs $n = 3$, the ground state is a gapless spin liquid.

In [139], a 1/3 plateau on the magnetization curve was predicted in numerical calculations for clusters of ladders with limited dimensions with three legs. Figure 3.1 shows the dependence $M(B)$ for the exchange ratio along the rung J' to the exchange along the guide J $J'/J = 3$. The existence of such a plateau was predicted for strong interactions in the $J'/J \leq 2$ rung. Theoretically, the possibility of the appearance of a gap in the antiferromagnetic ladder $S = 1/2$ with three legs was considered for quantum tubes, that is, in the presence of a strong frustrating interaction in the rung between all three chains [140].

For such antiferromagnetic ladders, an interesting possibility of a spin–Peierls transition at T_{SP} was also predicted for the formation of a quantum ground state. In the presence of strong interaction along the rung, such a ladder can be regarded as an effective Heisenberg chain of half-integer spins. In the case of strong interaction with a lattice below T_{SP}, it can go to the dimerized state along the legs, as shown in Fig. 3.2 [141].

Despite a noticeable number of theoretical papers devoted to the quantum ground state of the antiferromagnetic ladder of spins $S =$

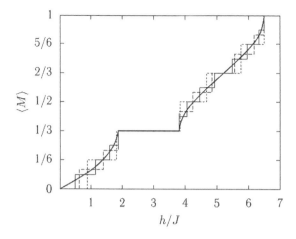

Fig. 3.1. Calculation of the magnetization curve for a ladder with three legs and the ratio of exchanges $J'/J = 3$. Thin dashed lines show calculations for clusters 4, 6 and 8 of links along the rung. The solid line shows the thermodynamic limit [139].

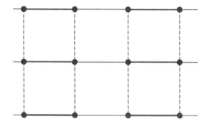

Fig. 3.2. Schematic representation of a state of the 'dimerized column' type in a ladder with three legs [141].

1/2, only a small number of compounds with such a topology of the magnetic subsystem have been found.

The only well-known representative of a flat ladder with three legs is $Sr_2Cu_3O_5$. In the orthorhombic crystal structure of this compound (space group I_{mmm}) copper cations Cu^{2+} are in a square oxygen environment. Squares of CuO_4, connected through vertices, form ladders with three legs, as shown in Fig. 3.3 (left). The temperature dependence of the magnetic susceptibility of this compound, shown in the right panel of Fig. 3.3, there is a decrease in $\chi(T)$ with decreasing temperature and a small increase at the lowest temperatures. The increase of $\chi(T)$ at low temperatures is associated with the presence of paramagnetic centres/defects in the sample. By

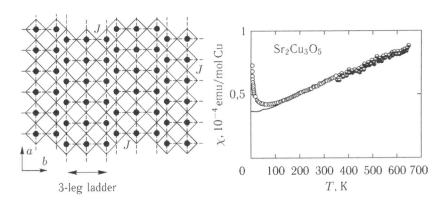

Fig. 3.3. The left panel: the crystalline structure of $Sr_2Cu_3O_5$ [142]. Right panel: temperature dependence of magnetic susceptibility. Open points are experimental data, the line is the result obtained after subtracting the contribution of defects.

subtracting this contribution from the experimental data, the magnetic susceptibility reached a constant value of $3.5 \cdot 10^{-5}$ emu/mol at the lowest temperature, which corresponds to the behaviour of a gapless spin liquid [142].

In the nuclear magnetic resonance experiment, the ^{63}Cu signal was not observed in $Sr_2Cu_3O_5$ below 100 K, which also indirectly indicates spin-liquid behaviour. A decrease in temperature can lead to an increase in the magnetic coherence length, which makes the decay rate of the spin response very small [142]. In studies using the nuclear quadrupole resonance method [143], it was shown that above 300 K the system behaves like a ladder with three legs, and below 300 K the exchange interactions between the ladders are triggered, which translates the system into a two-dimensional regime.

3.2. Spin ladders with an even number of legs: a spin liquid with an energy gap in the spectrum of magnetic excitations

The simplest example of a spin ladder with an even number of legs is a ladder with two legs, formed by Heisenberg ions with spin $S = 1/2$. The Hamiltonian of such a ladder with antiferromagnetic interactions along the rung and along the legs, J' and J, can be written as [144]:

$$\hat{H} = J \sum_{a=1,2} \sum_{i} \hat{S}_{i,a} \hat{S}_{i+1,a} + J' \sum_{i} \hat{S}_{i,1} \hat{S}_{i,2}, \qquad (3.1)$$

where $\hat{S}_{i,a}$ is the spin operator at site i of the ladder at rung a. The ratio of the exchange integrals J and J' is important for the formation of the ground state, since it affects the dispersion law $E(q)$ and the gap that separates the main nonmagnetic and first excited states.

If $J'/J \geq 1$, then the ground state of such a ladder can be represented as a set of isolated dimers composed of magnetic moments along the ranks. The total spin of such a ladder is zero at $T = 0$. The excited state of the ladder corresponds to the transition of one of the dimers to the excited triplet state $S = 1$. Exchange interactions along the legs serve to transfer the excitation along the ladder, which leads to the formation of a magnon zone with $S = 1$ described by the dispersion law:

$$E(q) = J' + J \cos(q). \tag{3.2}$$

The spin gap corresponds to the lowest excitation energy, which can be obtained for $q = \pi$, and thus $\Delta = J' - J$.

If $J'/J \geq 1$, the main singlet state of the ladder can be represented as dimers at each rung, bound by a weak antiferromagnetic coupling with dimers at neighbouring ranks (the model of the resonating valence bond) [145]. Then the dispersion law becomes:

$$E^2(q) = \Delta^2 + 4a\Delta(1 + \cos(q)), \tag{3.3}$$

where $a = \dfrac{1}{2}\dfrac{d^2 E}{dq^2}$ characterizes the propagation velocity of the spin wave. The minimum energy in this case corresponds to the value $q = \pi$.

In the limit $J'/J \ll 1$, the ladder can be represented as two chains not connected with each other. It was previously assumed that there is no gap in the excitation spectrum of such a ladder, as in the case of an isolated chain half-integer spins. In [145] it was shown, however, that the spin gap vanishes only for $J'/J = 0$ and has a finite value for any ratio $J'/J > 0$. In this case, the spin-wave band has a minimum at $q = \pi$ and a maximum at $q = \pi/2$ (for $J'/J = 0$) or $q = 0$ (for $J'/J = \infty$). The width of the band is determined by the antiferromagnetic interaction parameter in the J' level and varies from $\pi J/2$ (for $J'/J \to 0$) to $2J$ (for $J'/J \to \infty$). The value of the gap depends on the ratio of the exchange parameters J and J'. Thus, for comparable values of exchange by rung and directing $\Delta \approx 0.5J$ [138, 143, 146, 147].

The decrease in the ratio J'/J influences the dispersion law, which is characterized by a quadratic dependence near the minimum of the dispersion curve and linear sections for $J'/J \gg 1$ and $J'/J \ll 1$.

The correlation lengths of spin moments ξ are the same for an isolated chain and a spin ladder at high temperatures. With decreasing temperature ξ becomes larger in the ladder than in the chain due to the greater number of the nearest neighbours contributing to the correlation. Therefore, in the low-temperature region, the dispersion law for a ladder can be assumed to be quadratic.

The dependence $E(q)$ determines the behaviour of the susceptibility and specific heat. Thus, the temperature dependence of the magnetic susceptibility can be found through the dispersion law $E(q)$ using the expression [148]:

$$\chi = \beta \frac{z(\beta)}{1+3 \cdot z(\beta)}, \tag{3.4}$$

where

$$\beta = \frac{1}{T}, \quad z(\beta) = \frac{1}{\pi} \int e^{-E(q)} dq.$$

For the quadratic dispersion law, there are analytical temperature dependences of $\chi(T)$ and $C(T)$ at low temperatures ($T \ll \Delta$), which include the energy gap:

$$\chi(T) = \frac{1}{2\sqrt{\pi a T}} e^{-\Delta/T}, \tag{3.5}$$

$$C(T) = \frac{3}{4} \left(\frac{\Delta}{\pi a} \right)^{1/2} \left(\frac{T}{\Delta} \right)^{-3/2} \left[1 + \frac{T}{\Delta} + \frac{3}{4} \left(\frac{T}{\Delta} \right)^2 \right] \cdot e^{-\Delta/T}. \tag{3.6}$$

At high temperatures ($T \gg \Delta$), the magnetic susceptibility and specific heat can be used to estimate the exchange integrals in rung and along the ladder legs J' and J:

$$\chi(T) = \frac{1}{4T} - \frac{1}{8T^2} \left(J + \frac{J'}{2} \right) + \frac{3}{64T^3} JJ', \tag{3.7}$$

$$C(T) = \frac{3}{16T^2}\left(J^2 + \frac{JJ'}{2}\right). \tag{3.8}$$

It was shown that the spin gap is also present in the spin ladder with antiferromagnetic interaction along the legs ($J > 0$) and ferromagnetic interaction in rung ($J' < 0$) [149, 150]. For a weak ferro/antiferromagnetic bond according to the rung ($|J'|/J \ll 1$), the susceptibility is determined by the same expression (3.4). In the case of a strong ferromagnetic bond in rung, a ladder with two legs is equivalent to a chain of spins $S = 1$, bound by an antiferromagnetic interaction $J/2$ along the chain. An estimate of the gap for such a chain as $\Delta = 0.41J$ agrees well with the estimate for Haldane chains.

The motifs of spin ladders $S = 1/2$ were found in layered compounds based on Cu^{2+} and V^{4+}. In the absence of interactions between the layers, the long-range magnetic order at finite temperatures can not be achieved. However, the interactions between the ladders in the layer can substantially reduce the magnitude of the spin gap.

The $SrCu_2O_3$ compound has an orthorhombic crystal structure (the space group *Cmmm*), which is represented in the left panel of Fig. 3.4 [151]. Here, the Cu_2O_3 planes are separated by Sr^{2+} cations. Magnetoactive Cu^{2+} cations are in the square environment of O^{2-} ions. The overlapping of $d_{x^2-y^2}$ orbitals of copper ions with p_x and p_y orbitals of oxygen ions gives $180°$-superexchange along the metal–oxygen–metal path along the a and b axes, which according to the

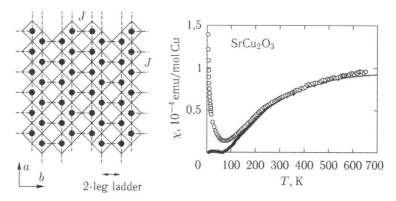

Fig. 3.4. Left panel: a fragment of the crystal structure of $SrCu_2O_3$. Cu^{2+} ions are represented in a square oxygen environment. Right panel: the temperature dependence of the magnetic susceptibility of $SrCu_2O_3$. The symbols are experimental data, the solid curve is the result of subtracting the van Vleck contribution, as well as the diamagnetic and impurity contributions [142].

Goodenough–Kanamori–Anderson rules will be antiferromagnetic. Thus, in the structure of $SrCu_2O_3$ it is possible to distinguish spin ladders with legs along the a axis and crossbars (ranks) along the b axis. The connection between the adjacent ladders corresponds to a weak 90° Cu–O–Cu ferromagnetic exchange. This exchange is further weakened by frustration due to the triangular arrangement of copper cations between the adjacent staircases. The temperature dependence of the magnetic susceptibility of $SrCu_2O_3$ is shown in the right panel of Fig. 3.4. The decrease of the $\chi(T)$ matrix (after subtraction of the impurity contribution) to zero at low temperatures indicates the singlet ground state of the system. The processing of the low-temperature region of the $\chi(T)$ dependence of the matrix made it possible to estimate the value of the spin gap by the formula (3.5) as $\Delta = 420$ K [142]. The estimate of the spin gap in an experiment on the study of nuclear magnetic resonance at ^{63}Cu was 680 K [152]. From data on inelastic neutron scattering, the spin gap in $SrCu_2O_3$ is $\Delta \approx 380$ K [153]. Estimates of the values of the exchange integrals along the guide and in rung also differ greatly depending on the method and make $J = 800$–2000 K and $J' = 750$–1000 K [154–156]. Apparently, this is due to difficulties in obtaining the $SrCu_2O_3$ phase without impurities.

3.3. Charge mechanism of dimerization of the spin ladder

In general, the antiferromagnetic spin ladders have a gap in the spectrum of magnetic excitations. However, the NaV_2O_5 compound is a rare example of the formation of the ground state in the ladder due to a phase transition at $T_C \sim 34$ K [157]. At this temperature, in NaV_2O_5 a redistribution of charges occurs between different positions of vanadium, which affects the crystal structure. As a result, a gap opens in the spectrum of magnetic excitations at T_C and a non-magnetic ground state is formed [157, 158].

At high temperatures, the tetragonal lattice NaV_2O_5 (the space group *Pmmn*) consists of two-dimensional layers connected along the edges and vertices of the pyramids VO_5. These layers are separated by Na atoms, as shown in Fig. 3.5. At high temperatures, all positions of vanadium are equivalent, and its formal oxidation state is +4.5 [159]. Consequently, one $3d$-electron is distributed between the two nearest vanadium ions, which form a rung of the spin ladder [160]. The spin ladders themselves are shifted half-way to each other along the b axis in the ab plane, as shown in Fig. 3.5. In the

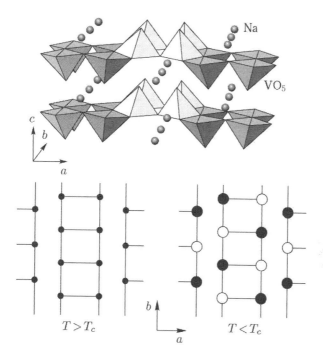

Fig. 3.5. Top panel: NaV_2O_5 crystal structure. The bottom panel: the structure of the spin ladders at $T > T_C$ and $T < T_C$. Magnetic V^{4+} and non-magnetic V^{5+} ions are shown with filled and empty symbols at $T < T_C$ [157].

standard spin ladder, there can be two differently directed spins on each site, therefore at high temperatures for NaV_2O_5 a ladder model is used that is filled by one quarter. This two-dimensional system of vanadium–oxygen planes can also be represented in the form of a set of non-interacting spin chains shifted by half a period relative to each other. The integral of the exchange magnetic interaction in the chain is estimated as $J = 280$ K [157].

At the phase transition temperature T_C, anomalies of all the physical properties of NaV_2O_5 are observed. The left panel of Fig. 3.6 shows the temperature dependence of the magnetic susceptibility of this system [157]. With decreasing temperature, the $\chi(T)$ dependence shows a wide maximum at 350 K, which corresponds to short-range correlation interactions in the chains. Above 150 K, the experimental dependence of $\chi(T)$ is well described by the Bonner–Fisher model for a homogeneous Heisenberg chain $S = 1/2$. Below 150 K, the experimental dependence of $\chi(T)$ deviates from the Bonner–Fisher model, which may be due to the development of spin fluctuations

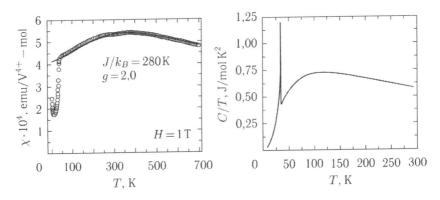

Fig. 3.6. Left panel: temperature dependence of the magnetic susceptibility of NaV_2O_5. The solid line shows a calculation using the Bonner–Fisher model in the interval 100–700 K [157]. The right panel: the temperature dependence of specific heat of NaV_2O_5 [161].

near the phase transition. Then at $T_C = 34$ K a sharp drop in the magnetic susceptibility is observed due to the appearance of a gap in the spectrum of magnetic excitations.

On the temperature dependence of specific heat $C(T)$ of NaV_2O_5, presented in the right panel of Fig. 3.6, the phase transition manifests itself as an sharp anomaly at T_C [161]. At temperatures below the structural transition, specific heat is well approximated by the sum of the lattice and magnon contributions: $C(T) = \beta T^3 + A_0 \exp(-\Delta/k_B T)$.

A feature of the structural phase transition in NaV_2O_5 is that the charge ordering of vanadium ions leads to the formation of a gap in the spectrum of magnetic excitations of this compound. At $T < T_C$, two non-equivalent vanadium positions appear in the structure. Moreover, the formal oxidation state of vanadium in these positions slightly deviates from the mean value of +4.5 and can be written as $V^{4.5-\delta}$ and $V^{4.5+\delta}$ (we can conditionally designate these positions as V^{4+} and V^{5+}) [162]. The pairs of magnetic V^{4+} and non-magnetic V^{5+} ions, connected by zigzag through the basal edges of the oxygen pyramids, form spin ladders, as shown in Fig. 3.5. Moreover, a separate spin ladder with such an arrangement of magnetic and non-magnetic ions does not contain a spin gap in the energy spectrum. However, the influence of the neighbouring ladders leads to an alternation of the exchange interaction and to the formation of a spin gap $\Delta = 114$ K in the spectrum of magnetic excitations of NaV_2O_5 according to theoretical [163] and experimental [164] studies.

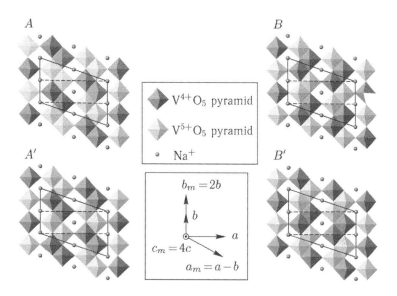

Fig. 3.7. Four variants of the arrangement of dimers (pairs of dark pyramids) formed by magnetic ions V^{4+}, for doubling the period in the *ab* layers of the NaV_2O_5 crystal structure at $T < T_C$ [166].

At $T < T_C$, a superstructure $(a{-}b) \times 2b \times 4c$ with monoclinic symmetry described by the C_2^3 space group is formed in NaV_2O_5 [165]. Due to the appearance of the non-equivalent positions of the ions V^{4+} and V^{5+}, the lattice period varies along the *a* and *b* axes ($a_m = a{-}b$, $b_m = 2b$). Figure 3.7 shows the device of a vanadium layer in a low-symmetry lattice. There are 4 variants of the arrangement of magnetic and non-magnetic dimers, which are denoted by the letters A, A', B and B'. To form a superstructure along the *c* axis ($c_m = 4c$), the layers should alternate in the order of ABA' B' [165] or AAA' A' [166]. According to the latest data, the structure of AAA' A' best allows us to describe the obtained X-ray diffraction spectra [166].

3.4. Combinations of spin chains and spin ladders

The $Sr_{14}Cu_{24}O_{41}$ metal oxide compound is characterized by orthorhombic symmetry (the space group Pcc_2) and contains in its structure alternating two-dimensional layers, one of which is made up of spin chains CuO_2, and the second – from Cu_2O_3 ladders, as shown in the upper panel of Fig. 3.8. These layers alternate along the *b* axis. The chains are somewhat displaced relative to each other in the *ac* plane, and the ladders are shifted relative to each other by a half-period along the *c* axis. The ratio of the periods of chains and ladders along the *c* axis can be represented as $10c_{chain} \approx 7c_{ladder}$. Thus 24 Cu^{2+} cations are distributed between the ladders and chains in the ratio of 14:10. In both the ladders and in the chains the Cu^{2+} cations are in a square oxygen environment. The distances between the copper ions along the legs and the rungs of the ladders are 1.90 Å and 1.97 Å, and along the chains 2.75 Å [167].

The magnetic susceptibility of $Sr_{14}Cu_{24}O_{41}$ contains the contributions of ladders, chains, Curie's contribution from defects/ impurities and the temperature-independent contribution of van Vleck [168]. On the $\chi(T)$ dependence shown in the bottom panel of Fig. 3.8 there is a wide maximum at 80 K, below which the magnetic

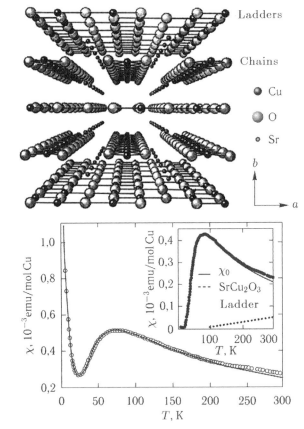

Fig. 3.8. Top panel: crystal structure of $Sr_{14}Cu_{24}O_{41}$. Bottom panel: the temperature dependence of the magnetic susceptibility of $Sr_{14}Cu_{24}O_{41}$. On the inset: the curve $\chi(T)$ after subtracting the paramagnetic and van Vleck's contributions. The dotted line shows the susceptibility of spin ladders [168].

susceptibility decreases noticeably. This behaviour may be due to the presence of spin ladders in the structure based on Cu^{2+} cations bound by strong antiferromagnetic exchange ($J \sim 1400$ K), which possess a singlet ground state. The magnitudes of the integrals of the exchange magnetic interaction along the legs and the ranks of the spin ladders are comparable with the corresponding parameters in $SrCu_2O_3$. At $T < 400$ K, the magnetic susceptibility is mainly determined by the contribution of copper chains.

The chains in $Sr_{14}Cu_{24}O_{41}$ contain not only Cu^{2+} magnetic ions, but also non-magnetic Cu^{3+} ions, alternating randomly at high temperatures. A decrease in the magnetic susceptibility below 80 K can be described in the dimer model (1.1) in the case when a part of the Cu^{2+} magnetic ions forms pairs separated by non-magnetic Cu^{3+} ions. For a good description of $\chi(T)$, the number of dimers from two spins $S = 1/2$ with an exchange magnetic interaction $J = 140$ K should be 1.47 per formula unit [168]. In nuclear magnetic resonance studies the charge ordering of Cu^{2+} and Cu^{3+} cations in chains was observed at temperatures below 80 K [169].

From the data on inelastic neutron scattering and diffraction of synchrotron radiation, two spin-density maxima are found which correspond to exchange interactions inside the dimers located in the chain [170, 171]. Moreover, the dimers are formed not by the nearest atoms, but by links located at a distance of 2 and 4 chain periods ($d_c = 5.48$ Å). Probably, these dimers are formed by two magnetic Cu^{2+} cations separated by Cu^{3+} non-magnetic cations. In the ground state of such a system, the dimers in the chains are ordered and the exchange interaction between the dimers J_c within the chain and between the J_a chains turned out to be close to $J_c \approx J_a \sim 10$ K.

3.5. Frame structures

S0ubstances related to the family of vanadium bronzes of the $\beta(\beta')$ type with the general formula $\beta(\beta')$-$A_xV_2O_5$ (A = Na, Sr, Cu, etc.) are quasi-one dimensional conductors [172]. One of the representatives of this family is a quasi-one dimensional compound with a variable valence β-$Na_{0.33}V_2O_5$, in which, with decreasing temperature, several order-disorder phase transitions are observed, each of which is associated with the ordering of one of the β-$Na_{0.33}V_2O_5$ subsystems. The substance subsequently undergoes structural ($T_S \sim 230$ K), charge ($T_C \sim 136$ K), and magnetic ($T_N \sim 22$ K) phase transitions [173–178]. In addition, under the action of a hydrostatic pressure of 8 GPa the charge-ordered phase in β-Na$_{0.33}$V$_2$O$_5$ collapses and a transition to the superconducting state occurs at a temperature of $T_{SC} = 8$ K [174].

The monoclinic crystal structure of this compound (the space group $A2/m$), characteristic of all vanadium bronzes in the $\beta(\beta')$ phase, contains tunnels elongated along the b axis formed by V–O complexes; Na ions are located inside the tunnels. At high temperatures in β-Na$_{0.33}$V$_2$O$_5$ for vanadium ions, there are three different crystallographic positions: V1 in the octahedral environment of oxygen ions forms zigzag chains along the z axis from the edge-shared octahedra VO$_6$, V2 in the same environment forms double chains of corner-shared octahedra, and V3 forms zigzag chains of coner-shared VO$_5$ pyramids, as shown in the upper panel of Fig. 3.9 [173]. In each unit cell there are also two crystallographic positions for Na$^+$ ions. With decreasing temperature, a structural ordering of Na ions occurs in the T_S, and a «superstructure» of the type $1 \times 2 \times 1$ appears along the b axis.

In the unit cell, β-Na$_{0.33}$V$_2$O$_5$, there is one V^{4+} ion ($S = 1/2$) and five nonmagnetic V^{5+} ions. At $T > T_C$, all vanadium positions are equivalent, and β-Na$_{0.33}$V$_2$O$_5$ is a conductor. As the temperature in this compound decreases, a metal-insulator transition takes place, such as charge ordering at T_C, nonmagnetic ions occupy positions V3, and positions V1 and V2 can be occupied by both V^{5+} and V^{4+}. According to the data of [175], the magnetic ions are in position V$_1$. The magnetic subsystem in Na$_{0.33}$V$_2$O$_5$ is highly dilute, since it contains magnetic and nonmagnetic ions in a ratio of 1: 5; however, with a further decrease in temperature, this substance observes the establishment of an oblique antiferromagnetic order below T_N, as shown in the lower panel of Fig. 3.9 [173].

Phase transitions are manifested in different ways in the temperature dependences of the specific heat, thermal conductivity $\kappa(T)$, resistivity $\rho(T)$, and thermal expansion coefficients $a_i(T)$ of this compound. The transition at $T_S \sim 230$ K, associated with the structural ordering of Na ions in VO tunnels, manifests itself as a weakly pronounced break in the $C(T)$ dependence and a change in the character of the $\rho(T)$ dependence. The metal–insulator transition at $T_C = 136$ K manifests itself in all the physical properties studied: as a peak on the $C(T)$ curve, as the change in the character of the temperature dependences $\kappa(T)$ and $\rho(T)$, and also as the peaks in the temperature dependences $a_i(T)$. The transition associated with the ordering of the magnetic subsystem at $T_N = 22$ K does not manifest itself in specific heat and thermal conductivity of

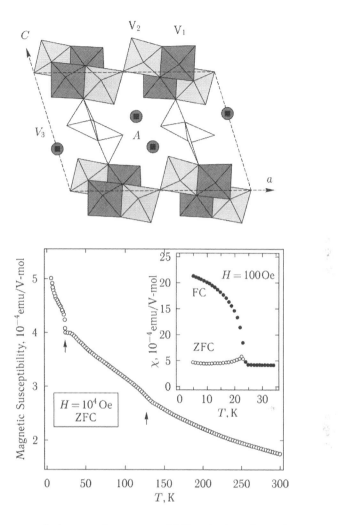

Fig. 3.9. The top panel: the crystal structure of β-Na$_{0.33}$V$_2$O$_5$, A is the double position for Na ions. The lower panel: the magnetic susceptibility of β-Na$_{0.33}$V$_2$O$_5$ [173]. The arrows indicate $T_C = 136$ K and $T_N = 22$ K. On the inset: the susceptibility measured in the ZFC and FC regimes.

β-Na$_{0.33}$V$_2$O$_5$, but is detected as anomalies of $\alpha_i(T)$. The fact that charge and magnetic orderings lead to the appearance of appreciable anomalies on the temperature dependences of the coefficients of thermal expansion testifies to the strong influence of the charge and magnetic subsystems on the lattice degrees of freedom in this compound.

The temperature dependence of specific heat $C(T)$ β-Na$_{0.33}$V$_2$O$_5$ is shown in the top panel of Fig. 3.10 [178]. A sharp peak of the

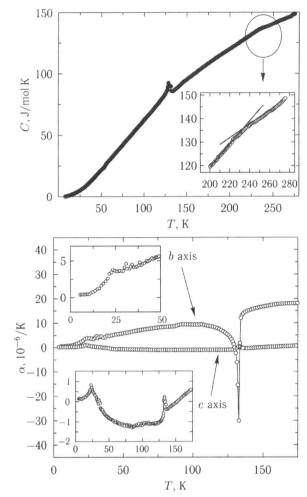

Fig. 3.10. The upper panel: the temperature dependence of specific heat of β-Na$_{0.33}$V$_2$O$_5$. Bottom panel: coefficients of thermal expansion of single crystal β-Na$_{0.33}$V$_2$O$_5$ $\alpha_i(T)$, along the axes b and c. The insertions show the behaviour of $\alpha b(T)$ in the vicinity of T_N and $\alpha_c(T)$ over a wide temperature range [178].

specific heat is observed at the temperature of the metal–insulator phase transition $T_C = 136$ K, and with a further increase in temperature there is a slight change in the slope of the $C(T)$ curve at a temperature of the structural phase transition $T_S \sim 230$ K, shown in more detail in the inset to the upper panel of Fig. 3.10. The absence of a noticeable anomaly specific heat of β-Na$_{0.33}$V$_2$O$_5$ at the magnetic ordering temperature $T_N \sim 22$ K can be explained as follows. In substances containing one-dimensional chains of magnetic ions bound by the exchange interaction, the establishment of a three-dimensional order in the entire magnetic subsystem is due to the presence of a weak interaction between these chains. The magnitude of this interaction and, accordingly, the ordering temperature are small in comparison with the magnitude of the interaction of the magnetic ions inside the chains. Thus, the magnetic entropy released at the temperature of three-dimensional ordering will be small and can not lead to the appearance of a significant anomaly in the temperature dependence of heat capacity. In addition, the magnetic subsystem β-$Na_{0.33}V_2O_5$ is highly dilute, and its contribution to specific heat is not very noticeable against the background of the rapid growth of the phonon contribution.

The temperature dependences of the thermal expansion coefficients $\alpha_i(T)$ of the β-$Na_{0.33}V_2O_5$ single crystal, studied along the crystallographic axes b and c, are shown in the bottom panel of Fig. 3.10 [178]. The coefficients of thermal expansion are strongly anisotropic. With increasing temperature, the single crystal β-$Na_{0.33}V_2O_5$ expands along the b axis and weakly shrinks along the c axis. As shown in the insets to the bottom panel of Fig. 3.10, noticeable anomalies of the positive sign along the two axes are observed at a magnetic ordering of T_N, which is caused by spontaneous compression of the single crystal at temperatures below T_N. Charge ordering with T_C is also accompanied by anomalies of the coefficients of thermal expansion along two directions, however, these anomalies differ in magnitude and sign. Along the b axis, there is a sharp anomaly with a negative sign, meaning that as the temperature decreases along the b axis the single crystal elongates. This behaviour is a sign of a first-order phase transition. An anomaly of the positive sign is observed along the c axis with T_C, indicating that the single crystal sharply decreases upon transition to a non-conducting state. The anomaly of the coefficient of thermal expansion along the c axis is much smaller than along the b axis, but the shape of the peak also indicates that the phase transition is a first-order transition.

Figure 3.11 shows the temperature dependence of the resistivity $\rho(T)$ of a β-$Na_{0.33}V_2O_5$ single crystal upon application of pressure [172]. The structural transition in T_S is manifested in this form in the form of a kink, and the metal–insulator transition with T_C is accompanied by a sharp increase in resistance. As the pressure increases, the metal–insulator transition is gradually blurred, and when the value of $P_C = 8$ GPa is reached, a sharp drop in the resistivity curve is observed, which indicates the transition of the

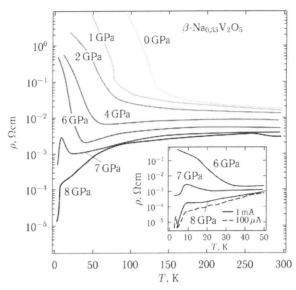

Fig. 3.11. Temperature dependences of the resistance of β-Na$_{0.33}$V$_2$O$_5$ single crystal along the b axis under pressure variation [172].

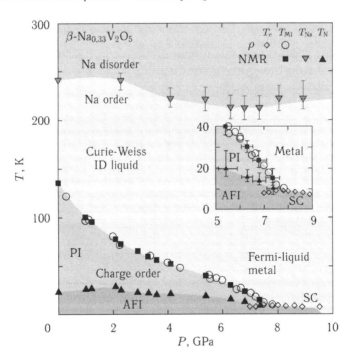

Fig. 3.12. The phase diagram of β-Na$_{0.33}$V$_2$O$_5$ in temperature–pressure coordinates [179]. Designations of areas: PI – paramagnetic insulator, AFI – antiferromagnetic insulator, SC – superconductor.

material to the superconducting state at $T_{sc} = 8$ K. These data are confirmed by studies of magnetic susceptibility under pressure.

Figure 3.12 shows the phase diagram of β-$Na_{0,33}V_2O_5$ in the temperature–pressure coordinates constructed as a result of the NMR study under pressure [179]. The antiferromagnetic order in the charge-ordered phase is suppressed with application of pressure, but it is present up to the lowest temperatures, while the substance is in the metallic state at a pressure less than the critical one. Measurements of the local spin density by the NMR method also make it possible to distinguish a smooth crossover from a one-dimensional Curie–Weiss spin liquid into a Fermi-liquid metal state when the critical pressure P_c is reached.

Quasi-two dimensional magnets with a square lattice

4.1. Quantum ground state

The interest in the study of quasi-two dimensional magnetic materials with a square lattice arose primarily in connection with the discovery of high-temperature cuprate superconductors $La_{2-x}Ba_xCuO_4$ [180]. The question of the quantum ground state of 2D magnetic materials with a square lattice is one of the most complicated, since in real substances the competition of intralayer exchange interactions, small interlayer exchange interactions, and anisotropy can radically affect the fundamental ordering mechanisms. This does not allow us to build a single model that could unambiguously predict the type of ground state. For a long time it was believed that in an ideal 2D magnet with a square lattice, the establishment of a long-range antiferromagnetic order is possible only at $T = 0$ [181]. Later it was shown in [182,183] that long-range order can exist in the ground state of an isotropic Heisenberg antiferromagnet on a square lattice for any $S \geq 1$. Numerical calculations show a reduction of the ordered spin component $\langle S_{zi} \rangle$ by about 40% [184], and this result agrees with the spin wave theory calculations involving $1/S$ corrections [185]. Rigorous proof of the existence or absence of long-range order in the ground state of an isotropic Heisenberg antiferromagnet on a square lattice is still not available, despite a large number of theoretical papers [183]. It was found experimentally that the spin order in 2D magnetic materials is extremely sensitive to the influence of an external magnetic field.

In this connection, it is interesting to mention the theoretical calculations of possible phases, including those induced by a magnetic field, in 2D Heisenberg magnets on a square lattice [186]. Interactions between the neighbors up to the third inclusive, J_1, J_2 and J_3 (Fig. 4.1) were taken into account in the model considered, and it was assumed that the exchange between the nearest neighbors J_1 is ferromagnetic, and the interactions between the second J_2 and third J_3 neighbours are antiferromagnetic. Calculations of the order parameter, performed by the Monte Carlo method, allowed to construct the phase diagram shown in Fig. 4.2. This figure shows three ordered phases analogous to the phases of 2D magnetic materials with a triangular lattice: a coplanar 120° spin order (lower phase), a phase with a 1/3

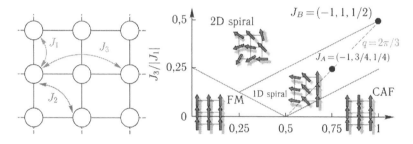

Fig. 4.1. 2D square lattice of spins, taking into account the interaction between neighbors up to the third inclusive, J_1, J_2 and J_3 (on the left) and the diagram of the basic quantum states depending on the ratio of J_1, J_2 and J_3 (right) [186].

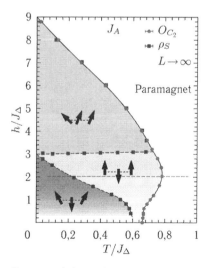

Fig. 4.2. The phase diagram of the main quantum states of a frustrated Heisenberg 2D magnet in an external magnetic field [186].

magnetization plateau with an up-up-down spin order (middle phase) and beveled order (upper phase).

4.2. The Berezinsky–Kosterlitz–Thouless transition

For 2D planar antiferromagnets Kosterlitz and Thouless, and independently of them, Berezinsky predicted the existence of a phase transition (BKT transition) associated with the formation of magnetic vortices on a frustrated square lattice (XY model) [187–190]. The formation of topological defects in the form of vortex-antivortex pairs, as shown in Fig. 4.3, occurs at a finite temperature $T_{KT} \neq 0$ because of the short range of action of the spin correlations, which leads to an additional degree of freedom characterized by the chirality vector.

In real 2D magnets, the long-range order is often established at a finite temperature T_N, which is quite high. This can not be explained only by a weak interlayer interaction. In this connection, an important influence of anisotropy is also taken into account in explaining the fundamental ordering mechanisms in 2D magnets [191]. It was shown [192,193] that even an arbitrary small easy-plane anisotropy breaks the isotropic 2D Heisenberg situation and can lead to a BKT transition at a finite temperature T_{KT}. Monte Carlo calculations predict that a quantum antiferromagnet with a square lattice of spins with small easy-plane anisotropy undergoes a crossover from high-temperature isotropic behaviour to the 2D XY behaviour at a temperature of ~30% higher than T_{KT} [192, 193]. A slightly higher T_{KT} realizes a long-range 3D ordering with T_N induced by the emerging intralayer BKT transition. It is predicted that a corresponding weak

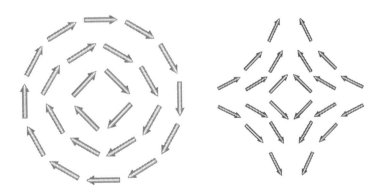

Fig. 4.3. The coupled vortex–antivortex pair at $T < T_{KT}$

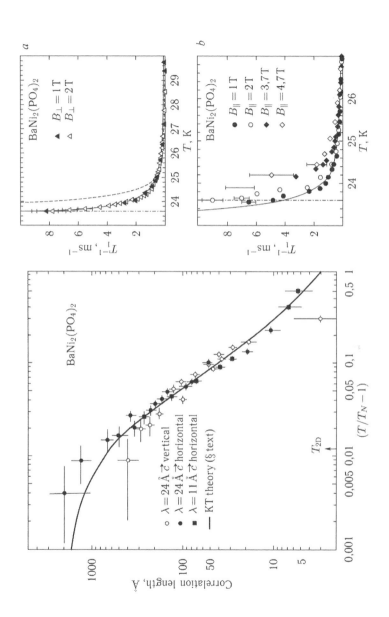

Fig. 4.4. Left panel: the temperature dependence of the spin correlation length obtained from the neutron scattering data of $BaNi_{12}(PO4)_2$. Right panel: *a*) the field perpendicular to *ab*; *b*) the field in the plane *ab*. The solid curves show the approximation of the experimental data in the BKT model.

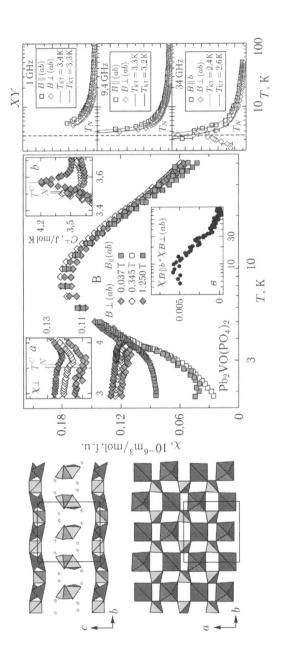

Fig. **4.5.** The crystal structure of the 2D magnet Pb$_2$VO(PO$_4$)$_2$ (left) [198], the temperature dependence of the magnetic susceptibility (in the middle), and the width of the ESR line (right) [197]. The solid curve shows the approximation in the BKT model.

anomaly in the specific heat should appear in a narrow temperature range in the vicinity of the T_{KT}. Moreover, it was shown that even in the case of an isotropic 2D Heisenberg antiferromagnet, the BKT transition can be induced by an external magnetic field, so that a truly 2D XY behaviour can be observed in a wide temperature range [194].

The experimental confirmations of the BKT transition are extremely few and mostly obtained by methods that allow us to extract information on the spin dynamics. In particular, the dynamics of the spin-correlation length $\xi(T)$ in accordance with the BKT transition scenario was found from experiments on elastic and inelastic neutron scattering for a 2D square antiferromagnet $BaNi_2(PO_4)_2$ with $T_N = 23.6$ K [195]. The solid curve in the left panel of Fig. 4.4 corresponds to the approximation of the experimental data in the temperature range $10^{-2} \leq T/T_N - 1 \leq 0.5$ by the formula

$$\xi(T) = A \exp B \left(\frac{T}{T_{KT}} - 1 \right)^{-1/2}.$$

The parameters obtained were: $T_{KT} = 0.96 T_N$, $A = 0.6$ Å and $B = 1.6$, which is in good agreement with the BKT theory, where $B_{theor} = \pi/2$. Similar studies for the isostructural arsenate $BaNi_2(AsO_4)_2$ yield BKT parameters of the model $T_{KT} = 0.95 T_N$, $A = 0.8$ Å and $B = 1.5$. These results give a direct indication of the realization of the BKT transition in these 2D magnets. Additional confirmation of the BKT transition with $T_{KT} = 0.95 T_N$ was obtained for $BaNi_2(PO_4)_2$ by NMR, as shown in the right panel of Fig. 4.4 [196].

ESR studies on a 2D $Pb_2VO(PO_4)_2$ antiferromagnet with a square (XY) lattice of V^{4+} ($S = 1/2$) lattice in a pyramidal oxygen environment also exhibit a spin dynamics in accordance with the formation of in-plane magnetic vortices at $T_{KT} = 0.85 T_N$ [197] , as shown in Fig. 4.5. Detailed studies of the width of the ESR line with variation of the magnetic field show that the critical temperature T_{KT} decreases with increasing external field strength.

4.3. Manganese chromate

The metastable phase of $MnCrO_4$ is characterized by an orthorhombic crystal lattice with the space group *Cmcm* [199]. The isostructural compounds $M^{2+}CrO_4$ (M = Mg, Cd, Ni, Co, Cu) with magnetic cations are well known as quasi-one dimensional magnets [200–203], with $CuCrO_4$ copper chromate being claimed as a possible multiferroic. In the high-pressure stable phase of $MnCrO_4$, the compound has a rutile structure. Its charge composition is $Mn^{5+}Cr^{3+}O_4$, all cations of manganese and chromium are in the octahedral coordination. In the metastable phase of $MnCrO_4$, chromium appears in a tetrahedral oxygen environment, providing a charge balance of $Mn^{2+}Cr^{6+}O_4$; Thus, magnetism is determined only by manganese ions, which form chains of edge-shared MnO_6 octahedra along the [001] axis (Fig. 4.6). Along two other directions, the chains are connected by non-magnetic tetrahedrons $Cr^{6+}O_4$, so that the structural conditions allow expecting weak interaction between $S = 5/2$ spin chains.

The temperature dependences of the static magnetic susceptibility in the field $B = 0.1$ T, shown in Fig. 4.7, exhibit non-trivial behaviour at low temperatures. The most striking feature is the appearance of a broad maximum at $T_{max} \approx 42$ K, which indicates either a low-dimensional correlation maximum or a phase transition to an ordered state. Analysis of the dependence of $\chi(T)$ in comparison with its first derivative $d\chi/dT(T)$ reveals an additional weakly pronounced anomaly at $T_2 \sim 9$ K, as seen in the upper inset to Fig. 4.7. At the lowest temperatures ($T < T_2$) under the FC regime, the magnetic susceptibility again increases, which may be due to the presence of a small amount of impurity in the sample under study. Below T_{max}, there is a strong discrepancy between the ZFC and FC dependences

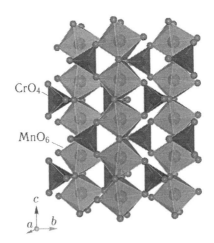

Fig. 4.6. The polyhedral form of the crystal structure of $MnCrO_4$ [199].

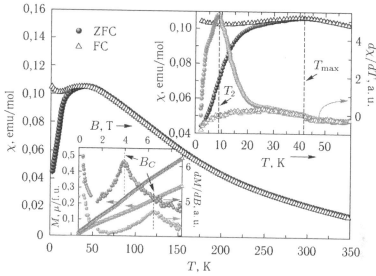

Fig. 4.7. Temperature dependences of the magnetic susceptibility of MnCrO$_4$. Top insert: enlarged low temperature area. Lower insert: $M(B)$ and $dM/dB(B)$ dependences at $T = 2$ and 5 K.

$\chi(T)$, indicating possible spin-glass behaviour at low temperatures. In contrast to the isostructural analogue of CuCrO$_4$ [200], the temperature dependence of the magnetic susceptibility $\chi(T)$ MnCrO$_4$ does not obey the Curie–Weiss law up to the highest temperature achieved in the experiment of ~ 350 K.

The bottom inset in Fig. 4.7 shows the magnetization isotherms $M(B)$ measured at low temperatures, which do not show saturation and hysteresis in the investigated range of magnetic fields up to 9 T. The maximum magnetic moment remains significantly lower than the the saturation moment $M_s \approx 5\mu_B$ per ion Mn for the Mn^{2+} ion ($S = 5/2$). At the same time, a change in the curvature of the field dependences $M(B)$ is revealed, which is clearly manifested in the form of wide maxima on the derivatives with respect to the field $dM/dB(B)$. This behaviour corresponds to the appearance of a spin-reorientation transition induced by a magnetic field. The position of the critical field estimated from $dM/dB(B)$ is $B_{SF} \sim 4$ T at $T = 2$ K. It should be noted that the observed feature is very broad, which is unusual for classical 3D antiferromagnets and can correspond to the phase diagram regions at low temperatures, where the long-range order is replaced by the short-range one.

The temperature dependences of the magnetic susceptibility of MnCrO$_4$ were analyzed in the framework of two models: a 1D antiferromagnetic chain and a 2D square lattice of spins $S = 5/2$. None of these models provided satisfactory agreement with the experimental data. Thus, unlike the isostructural analogue of $CuCrO_4$ [200], the magnetism of $MnCrO_4$ can not be described in the framework of the 1D antiferromagnet model.

Direct confirmation of the establishment of the long-range order was obtained from data on the specific heat of $MnCrO_4$, as shown in Fig. 4.8, which reveal an anomaly of the λ-type at $T_N \approx 42$ K, which is characteristic of the second-order phase transition to the 3D magnetically ordered state. The data on the specific heat are also confirmed by the presence of a second anomaly at $T_2 \sim 9$ K, which manifested itself in the $\chi(T)$ dependence. This anomaly is seen in the representation in the coordinates C/T from T on the upper inset to Fig. 4.8. Figure 4.8 shows the specific heat of the isostructural non-magnetic analogue $InVO_4$, which is used to estimate the lattice contribution to the specific heat of $MnCrO_4$. The magnetic contribution to the specific heat of $MnCrO_4$ C_m was obtained by subtracting the lattice contribution from the experimental values of the total specific heat and is shown in the lower inset to Fig. 4.8. An analysis of $C_m(T)$ below T_2 within the framework of the theory of spin waves in accordance with the power law gives, with good accuracy, the values of $d = 2$ and $n = 1$, which implies the presence of 2D AFM magnons at low temperatures. However, as mentioned above, the $\chi(T)$ analysis within the 2D square lattice model does not allow an adequate description of the experimental data.

As can be seen on the upper inset to Fig. 4.8, the magnetic entropy of $MnCrO_4$ saturates at a level of ~14 J/mol K, which is close to the value from the estimate for the mean-field theory $\Delta S_m = R\ln(2S + 1) \approx 15$ J/mol K. Below the T_N temperature, about 60% of the entropy is released, which indicates a large contribution of short-range correlations, characteristic of low-dimensional and frustrated magnetic systems [191].

From an analysis of the crystal structure it follows that there are four possible ways of exchanges, shown in Fig. 4.9: it is necessary to take into account two parameters J_{nn} and J_{nnn} of interacting spins in the chains of the edge-shared MnO_6 octahedra along the c axis, as well as two interchain exchange parameters J_1 and J_2. The results of the first-principles calculations are presented in Table 4.1. As can be seen, the dominant exchange parameters J_{nn} and J_1 are antiferromagnetic and comparable in magnitude. In this case, J_{nn} corresponds to the super-superexchange interaction Mn–O–Mn, while

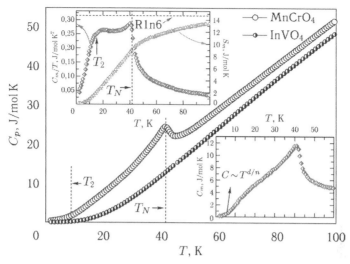

Fig. 4.8. Temperature dependences of the specific heat for $MnCrO_4$ and its non-magnetic analogue $InVO_4$ in a zero magnetic field. On the insets: the magnetic specific heat and magnetic entropy for $MnCrO_4$ (above) and the increased low-temperature part and its approximation in accordance with the power law for magnons $C_m \sim T^{d/n}$ (below).

Table 4.1. The parameters of the exchange integrals obtained from the calculations by the GGA + U method and the Curie–Weiss temperature (in Kelvins) for $MnCrO_4$

	$U_{eff} = 3$ eV	$U_{eff} = 4$ eV	$U_{eff} = 4$ eV
J_{nn}	−2.7	−2.2	−1.9
J_{nnn}	0.1	0.0	0.0
J_1	−2.7	−2.2	−1.8
J_2	−0.1	−0.1	−0.1
Θ	−95.4	−77.1	−63.6

J_1 corresponds to the supersuper exchange interaction Mn–O...O–Mn. In spite of the fact that the supersuper exchange interaction is often neglected, it may prove to be stronger than direct super-superexchange interaction [204, 205]. Each high-spin ion Mn^{2+} ($S = 5/2$) has five orbitals on the d-shell, filled by 1/2, i.e., all t_{2g} and e_g electronic orbitals take part in the exchange. If the oxygen atoms in each MnO_6 octahedron are classified into basal O_{eq} and apical O_{ap}, then from the structure analysis it is clear that the basal O_{eq} are common for edge-shared MnO_6 chains. Thus, each octahedron MnO_6 has four Mn–O_{eq} bonds and two Mn–O_{ap} bonds. The MnO_6 octahedra in adjacent chains on the J_1 exchange path are connected

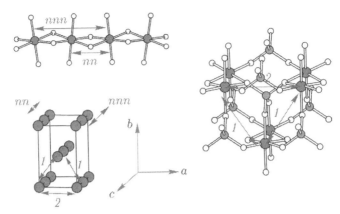

Fig. 4.9. Fragment of the crystal structure and the main paths of exchange interactions for $MnCrO_4$. The Mn, Cr, and O atoms are shown in successively lighter shades, respectively.

by means of two $O_{ap} \ldots O_{eq}$ contacts, whereas on the J_2 exchange path, the connection is carried out with the participation of two $O_{eq} \ldots O_{eq}$ contacts. In this case, the cation–anionic bonds of $Mn-O_{ap}$ are longer than the $Mn-O_{eq}$ bonds (2.173 vs 2.110 Å), and therefore the bond length $O_{eq} \ldots O_{eq}$ is larger than $O_{ap} \ldots O_{eq}$ (2.752 vs. 2.714 Å), which agrees well with the result obtained $J_1 \gg J_2$. It is interesting to note that, unlike the picture described above for $MnCrO_4$, in its isostructural analogue $CuCrO_4$, each Cu^{2+} ion ($S = 1/2$) has only one half-filled orbit in the plane of edge-bonded octahedra CuO_6, which leads to a very weak interchain interaction and causes 1D magnetism of copper oxide.

Thus, with the dominance of J_{nn} and J_1 exchanges, the magnetic sublattice should be considered as square planes in the [110] direction (they are not square in geometry, however, they are square in terms of magnetic interactions, since $J_{nn} \sim J_1$). These planes of squares are added so that the ions in adjacent layers fall into the middle of the square of the adjacent layer. Thus, the interplanar exchange turns out to be frustrated.

Thus, the quantum ground state for $MnCrO_4$ is the antiferromagnetic state, and the temperature of the maximum $T_{max} \approx 42$ K on $\chi(T)$ should be considered as the Néel temperature, in spite of the absence of the sharp maximum on the susceptibility typical of the 3D antiferromagnetic ordering.

The evolution of the ESR spectra with a temperature variation for the $MnCrO_4$ powder sample is shown in Fig. 4.10. The shape

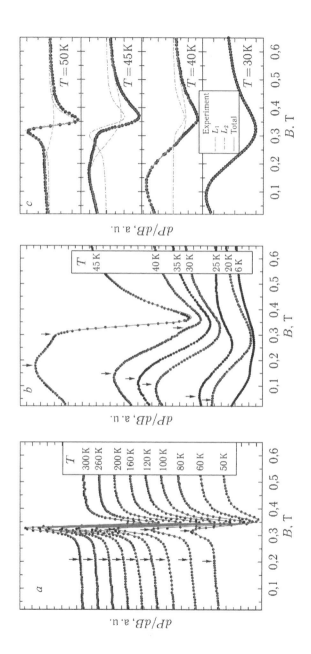

Fig. 4.10. The temperature evolution of the first derivative of the absorption line for MnCrO₄ (*a, b*) and an example of the expansion of the ESR spectrum into two resolved components (*c*): the points are experimental data, the lines are approximations.

of the ESR lines of the spectrum changes sharply with decreasing temperature: distortion and splitting of the line appear, most noticeable at temperatures below ~70 K. Thus, two distinct resonance modes are seen in the low-temperature region. The temperature behaviour of these two components is different. The resonant field of the narrower and less intense L_2 mode changes little wth decreasing temperature, while the broad and intense L_1 line, which makes a decisive contribution to the absorption in the entire region of the investigated temperatures, shifts noticeably toward lower fields at $T < 70$ K (Fig. 4.10 *b*). The linewidth also demonstrates a brighter temperature dependence for the L_1 mode. The ESR signal is degraded in the vicinity of the Néel temperature, which is in agreement with the establishment of the long-range order and, as a consequence, the opening of a gap in the spectrum of magnetic excitations.

The analysis of the shape of the ESR line was carried out taking into account the fact that the sample under study is a semiconductor, which makes it possible to use the Dyson type profile for analyzing the line shape. In addition, the experimental absorption line is relatively wide (only one order of magnitude smaller than the resonance field in a given compound). In this case, the approximating formula includes two circular components of a linearly polarized high-frequency field:

$$\frac{dP}{dB} \propto \frac{d}{dB}\left[\frac{\Delta B + \alpha(B - B_r)}{(B - B_r)^2 + \Delta B^2} + \frac{\Delta B - \alpha(B + B_r)}{(B + B_r)^2 + \Delta B^2}\right], \qquad (4.1)$$

where α is the asymmetry parameter that determines the fraction of the dispersion in the absorption spectra, B is the magnetic field, B_r is the resonant field, and ΔB is the line width. It can be seen that the lines of approximation (solid lines in Fig. 4.10) are in good agreement with the experimental data. An example of the expansion of the spectrum into two lines for several temperatures is shown in Fig. 4.10 *c* (dashed lines – 2 components in this order, solid line – the sum of these components).

The values of the effective *g*-factors proved to be practically isotropic for both components and averaged $g_1 = 2.05 \pm 0.03$ for the main (broad) line L_1 and $g_2 = 1.97 \pm 0.01$ for the narrow line L_2. As the temperature decreases g_2 remains practically unchanged, while g_1 strongly increases below ~75 K. The width of the narrow line ΔB_2 is also practically independent of temperature and exhibits a sharp increase only in the immediate vicinity of T_N, on the contrary, the broadening of the L_1 line occurs already at temperatures of ~180 K. Such behaviour indicates the presence of an extended area and short-range order correlations at temperatures much higher than T_N, which is characteristic of low-dimensional magnetic systems. Similar to the behaviour of static magnetic susceptibility, the integrated intensity of ESR spectra (proportional to the number of magnetic spins) for monotonically increasing resonance modes increases monotonically with decreasing temperature, passes through a maximum at 40 K, and then decreases. Unlike the static magnetic susceptibility, the temperature dependence of the dynamic susceptibility $\chi_{esr}(T)$ of each individual component follows satisfactorily the Curie–Weiss law in the paramagnetic region.

The results obtained with respect to the spin dynamics can naturally be interpreted under the assumption that two different paramagnetic centres coexist. It is logical to assume that the main resonant mode L_1, with a characteristic value of the effective g-factor of $g_1 \approx 2.05$, corresponds to the main signal from Mn^{2+} ions in a distorted octahedral coordination with a g-factor value typical of S-ions close to $g \sim 2$. This signal dominates over the entire temperature range and, apparently, reflects the key role of Mn^{2+} ions in the magnetism of $MnCrO_4$. It can also be assumed that the second, narrow and low-intensity component with $g \approx 1.97$ corresponds to the signal from a small number of defective paramagnetic centres that can be associated with the presence of Cr^{5+} ions. In this case, naturally, one should expect that due to a change in the valency of some of the Cr ions from 6+ to 5+, the corresponding part of the Mn ions should change its valence state from 2+ to 3+. It is difficult to distinguish in the ESR spectra the presence of Mn in such diverse states since the signals from the Mn^{2+} and Mn^{3+} ions have close g-factor values. We note that the indirect indication of the presence of Mn^{3+} ions follows from a significant broadening of the absorption line, which is much broader for the Jahn–Teller ion of Mn^{3+}.

Confirmation of this hypothesis was obtained in the study of X-ray absorption spectra XANES. Simulation of the experimental absorption spectra showed that the best description of the spectra near the K-edge of absorption on Mn ions and on Cr ions is achieved on the assumption of the presence of a small concentration of the cations Mn^{2+}/Mn^{3+} and Cr^{5+}/Cr^{6+} with different valencies. Such disorder associated with partial charge transfer is naturally expected for a metastable $MnCrO_4$ compound synthesized at sufficiently low temperatures.

Quasi-two dimensional magnetics with a triangular lattice

5.1. Geometrical frustration

The most important characteristic of two-dimensional systems based on triangular geometry is the frustration of the magnetic subsystem. If a Néel antiferromagnetic state is realized on a square lattice under the conditions of antiferromagnetic interaction between the nearest neighbours, then in the case of a triangular lattice this is not so. If the exchanges between the nearest neighbours have an antiferromagnetic nature and are also equal or comparable in magnitude, then the interaction of ion 1 with ion 2 can interfere with interactions 1–3 and 2–3, as shown in Fig. 5.1. If the spin has only two possible locations – up or down (the case of strong anisotropy – the Ising model), then simultaneous minimization of energy for all pair interactions is impossible. As a result of frustration, the effective exchange interaction is strongly attenuated, and the magnetic subsystem experiences difficulties in the formation of the

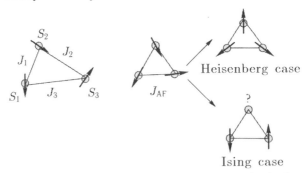

Fig. 5.1. Geometrical frustration on a triangular lattice.

long-range magnetic order. The Néel order of an antiferromagnet on a triangular lattice is not energy-minimizing, even in the classical limit.

In accordance with the Mermin–Wagner theorem [181], for any non-zero temperature, the one or two-dimensional Heisenberg system can not experience either a ferromagnetic or an antiferromagnetic ordering.

The ground state of the two-dimensional triangular Heisenberg lattice with $S = 1/2$ is shown in Fig. 5.2. All the magnetic moments are related by the same antiferromagnetic frustrated exchange J and whose energy is described by the Hamiltonian [206]

$$\hat{H} = J \sum_{\langle i,j \rangle} \vec{S}_i \cdot \vec{S}_j + ..., \qquad (5.1)$$

The magnetic moments in the structure are located in the plane at an angle of 120° with respect to each other. Such an ordered state does not have a gap in the spectrum of magnetic excitations. This structure of the arrangement of magnetic moments was first proposed by Yafet and Kittel in [207] to describe the antiferromagnetic state in ferrites.

The presence of a 1/3 plateau is predicted on the magnetization curve of two-dimensional Heisenberg antiferromagnets on a triangular lattice, as shown in the left panel of Fig. 5.3, which corresponds to the collinear arrangement of the three sublattices, which is stabilized by thermal and quantum fluctuations [208,209]. In an external magnetic field, the fluctuations lead to two new phases: the coplanar three-sublattice 'Y phase', which is a 'canted' version of

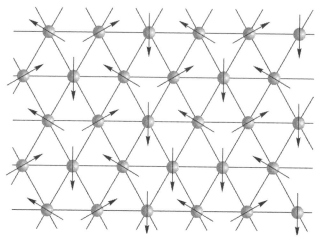

Fig. 5.2. The ground state of a Heisenberg antiferromagnet on a triangular lattice [206].

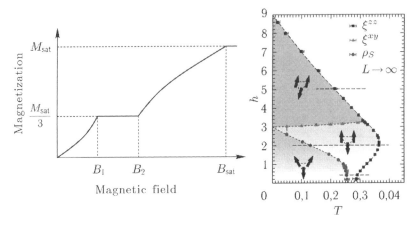

Fig. 5.3. Left panel: the theoretical form of the magnetization curve of a two-dimensional Heisenberg antiferromagnet with a triangular lattice [209, 210]. Right panel: magnetic phase diagram.

120°, the state and the 'unfolded' 2: 1 phase, as shown in the right panel of Fig. 5.3.

From the calculation of the field dependence of the magnetization by the Monte Carlo method on clusters [211] for the Hamiltonian:

$$\hat{H} = \sum_{\langle i,j \rangle} \vec{S}_i \cdot \vec{S}_j - \lambda \sum_i S_i^z, \qquad (5.2)$$

where λ corresponds to the strength of the external magnetic field, it was shown that the total saturation on the magnetization curve is achieved at $\lambda_S = 4.5$. The plateau of the magnetization 1/3 is present in the field interval $1.3 < \lambda < 2$.

In the Ising case, the magnetic order is absent down to low temperatures, and disorder remains in the ground state. The measure of disorder can be residual entropy, which makes the system with a triangular lattice a candidate for realization of the spin-liquid state [212]. According to theoretical estimates, in the triangular lattice the residual entropy is: $\Delta S^0 = 0.323 k_B N$, where N is the number of magnetic moments. For a triangular lattice, a model of resonant valence bonds was proposed [213], the ground state in which is a linear superposition of a large number of electronic singlets, and elementary excitations are transmitted by spinons with $S = 1/2$, as shown in the right panel of Fig. 5.4. Such a ground state at $T = 0$ is disordered and is an example of a spin liquid.

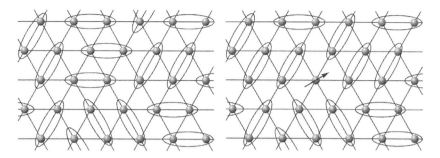

Fig. 5.4. The main (left panel) and the excited (right panel) state of the resonating valence bonds for the triangular lattice [206].

Detailed theoretical calculations of the effect of an external magnetic field on the magnetic phase diagram of triangular antiferromagnets with spin $S = 1/2$ were performed in [213]. As a model object, three homogeneous chains with interchain exchange J and interchain diagonal exchange J' were taken, as shown in the upper panel of Fig. 5.5, so that on the whole the lattice can be represented as triangular spin tubes. The phase diagram obtained in this case is shown in the lower panel of Fig. 5.5 in the axes: the magnetic field – the anisotropy degree $R = 1 - J'/J$ is shown in the lower panel of Fig. 5.5. The phase diagram obtained dominates the quantum states, and the classical state is realized only in a small region. Among the phases generated by quantum effects there are commensurate and incommensurate coplanar quasi-ordered states that appear in the immediate vicinity of the isotropic region for most fields, as well as in the high-field region for most anisotropic states (C – commensurable phase, IC – incommensurate phase, SDW – spin density wave). Dotted lines in the phase diagram correspond to constant values of magnetization at the level of 5/6, 1/2 and 1/6 of the saturation moment for the upper, middle and lower lines, respectively. In the region of weak fields, the dimerized phase is realized, which is largely due to the appearance of a one-dimensional character of the correlations in triangular spin tubes. It is interesting that the largest region of the phase space is occupied by the phase of the spin density wave, which has incommensurate collinear correlations along the field. This phase has no classical analogue and can be related to the important role of one-dimensional fluctuations induced by frustration. The central cross section of the phase diagram corresponds to a phase with a plateau 1/3 of saturation magnetization with an energy gap for all excitations and a spin order of the 'up-up-down' type. Most of the above features are expected to be transferred to a purely two-dimensional system.

5.1.1. Lithium–nickel tellurate

The suppression of the magnetic order caused by frustration was observed in the quasi-2D magnet of Li_4NiTeO_6 [215]. The structure of the lithium–nickel layered tellurate is shown in Fig. 5.6. the magnetoactive mixed layers of cations in the octahedral oxygen environment alternate with non-magnetic layers of lithium. In the magnetoactive layer there is practically ideal cation ordering, so that around the TeO_6 octahedron there are three nickel ions and three lithium ions forming a hexagonal cell from the edge-shared alternating octahedra NiO_6 and LiO_6, as shown in the right panel of Fig. 5.6. Thus, the triangular magnetic subsystem of Ni^{2+} ions is realized when half of the magnetic ions in the honeycomb layers $(LiNiTeO_6)^{3-}$ are replaced by non-magnetic lithium cations. Direct

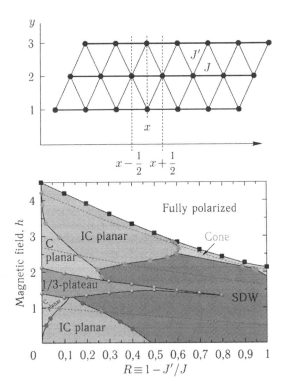

Fig. 5.5. The representation of a 2D triangular lattice with $S = 1/2$ as three interacting chains with interchain exchange J and interchain exchange J' (top) and a phase diagram of various basic quantum states in an external magnetic field (bottom) [213].

contacts of the NiO_6 octahedra are absent, and as a basic exchange interaction a weak supersuper-exchange with the participation of non-magnetic Li^+ and/or Te^{6+} cations is expected, as well as probable frustration of the magnetic subsystem based on the triangular geometry.

The temperature dependences of the magnetic susceptibility, shown in Fig. 5.7, do not show any discrepancies in the ZFC and FC modes, and in the whole investigated temperature range show a monotonous increase with cooling. This indicates that there is no long-range magnetic order up to a temperature of 1.8 K. In this case field dependences have a characteristic S-shape, as shown in the inset to Fig. 5.7, which indicates a significant role of frustration. In the temperature range above 100 K, the temperature dependence of the magnetic susceptibility is satisfactorily described by the Curie–Weiss law. The Weiss temperature is negative and equal to $\Theta \sim -11.4$ K, which indicates the dominance of antiferromagnetic exchanges and significant frustration. Assuming a magnetic ordering temperature below 1.8 K, we obtain an estimate for the frustration parameter $f = \Theta/T_N > 6$. The effective magnetic moment corresponds to the high-spin state of the Ni^{2+} ion.

In the ESR spectra, a single broadened Dyson line is observed, which indicates that the dispersion plays an important role in absorption, therefore, formula (4.1) was used to analyze the shape of the line. The results of the approximation are shown by solid lines in the left panel of Fig. 5.8. As can be seen, the theoretical curves describe the experimental data well. The spectral parameters obtained from the analysis are shown in the right panel of Fig. 5.8.

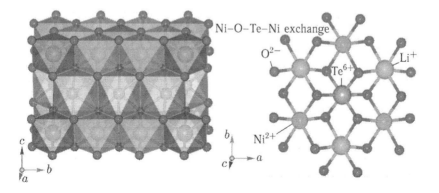

Fig. 5.6. The crystal structure of Li_4NiTeO_6: a general polyhedral species (left); a fragment of the $C2/m$ structure in the ab plane, showing the triangular arrangement of Ni^{2+} cations in the magnetoactive layer (on the right).

The average effective g-factor is $g = 2.08$ at room temperature that is typical of the Ni^{2+} ion in the octahedral oxygen coordination. The g-factor somewhat decreases with decreasing temperature, but the shift of the resonance field most likely corresponds to an increasing error in the determination, related to a significant line broadening and a significant asymmetry of the absorption line, as seen in the upper panel of Fig. 5.8 for $\alpha = 0.3-0.5$.

The absorption line is significantly broadened with decreasing temperature. An analysis of the broadening of the ESR line was carried out within the framework of the Mori–Kawasaki–Huber theory [102–106] in accordance with expression (2.15). The estimates obtained for the possible ordering temperature $T_N^{ESR} \sim 1.8$ K and the critical exponent $\beta = 1.59$ indicate about the strongly frustrated and two-dimensional character of the magnetic correlations in the investigated sample [216, 217]. The fact that a significant line broadening begins even at temperatures above 100 K indicates the presence of strong short-range magnetic correlations over a wide range of temperatures, which is characteristic of compounds with a reduced dimension of the magnetic subsystem [191]. It should be noted that the obtained value of the critical exponent is noticeably higher than the characteristic values for the

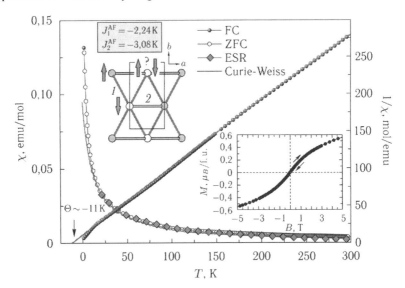

Fig. 5.7. Temperature dependences of magnetic susceptibility and integral ESR intensity for Li_4NiTeO_6. In the insets: the main paths of exchange interactions on the triangular lattice of nickel ions (top) and the field dependence of the magnetization at 2 K (bottom).

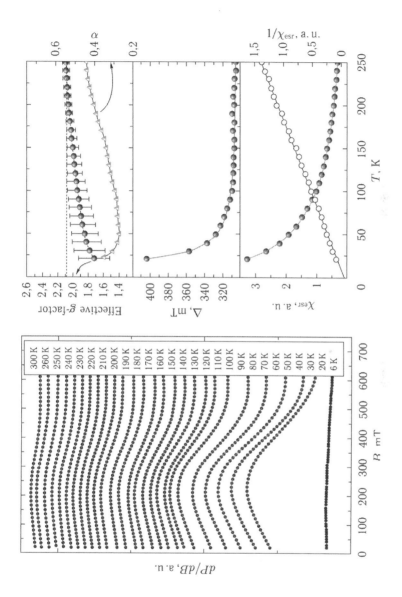

Fig. 5.8. Evolution of the ESR spectra for Li_4NiTeO_6 at a temperature variation (on the left) and temperature dependences of the effective g-factor, asymmetry parameter (α), ESR line width and ESR integral intensity (right). Points – experimental data, lines – the result of approximation.

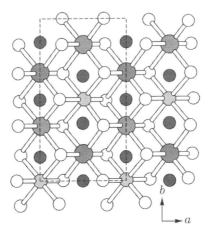

Fig. 5.9. The projection of an isolated layer $(LiNiTeO_6)^{3-}$.

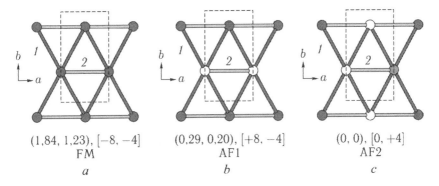

$(1.84, 1.23), [-8, -4]$	$(0.29, 0.20), [+8, -4]$	$(0, 0), [0, +4]$
FM	AF1	AF2
a	*b*	*c*

Fig. 5.10. Three possible spin configurations in an isolated layer $(LiNiTeO_6)^{3-}$. The numbers indicate the paths of the exchange integrals J_1 and J_2.

3D systems (0.5)–(1.2) and is well comparable to the theoretically expected value for 2D magnets $\beta = 3/2$ [216, 217].

Calculations from the first principles were performed to determine the nature of the exchange interactions. Taking into account the cation–cation distances, the two main paths of magnetic exchanges J_1 and J_2, from the crystal structure can be distinguished within each isolated $LiNiTeO_6$ layer, which form a triangular spin lattice in the investigated tellurate Li_4NiTeO_6 shown in Fig. 5.9.

To determine the values of the exchange integrals, we considered the three main spin-configuration states FM, AFM1 and AFM2, shown in Fig. 5.10. Empty and filled spheres correspond to the opposite orientation of the spins relative to the plane of the figure. The energy calculations for these spin configurations are performed and the

Table 5.1. The parameters of the exchange integrals and the Curie–Weiss temperature values obtained from the calculations by the DFT+U method for Li_4NiTeO_6

	$U = 3$ eV	$U = 4$ eV
J_1, K	-2.24	-1.49
J_2, K	-3.08	-2.06
Θ, K	-10.1	-6.7

obtained values of the exchange integrals J_1 and J_2, are presented in Table 5.1. In the framework of the theory of the molecular field [47], the Weiss temperature is related to the spin exchange parameters:

$$\Theta_{CW} = \frac{S(S+1)}{3k_B}\sum_i z_i J_i \approx \frac{4(2J_1 + J_2)}{3k_B}, \tag{5.3}$$

where the summation is over all nearest neighbors for a given spin site, z_i is the number of the nearest neighbors connected by the exchange integral J_i, S is the spin value ($S = 1$ for Ni^{2+}). The calculated values of Θ_{CW} are also given in Table 5.1. It can be seen that the experimental value of $\Theta = -11.4$ K agrees well with the theoretical value for the exchange parameters obtained with $U^{\text{eff}} = 3.0$ eV. As can be seen from Table 5.1, both exchange integrals are antiferromagnetic for both $U^{\text{eff}} = 3.0$ eV and for 4 eV. Thus, the spin subsystem of each layer of $LiNiTeO_6$ is strongly frustrated. This result agrees with the experimentally observed absence of the long-range magnetic order in Li_4NiTeO_6 up to 1.8 K.

5.1.2. Barium–cobalt antimonide

Barium–cobalt antimonide $Ba_3CoSb_2O_9$ is an example of a system based on a triangular geometry with $S = 1/2$, in which a plateau induced by a magnetic field is realized on the magnetization curve [218, 219]. The hexagonal crystal structure of $Ba_3CoSb_2O_9$ (the space group $P6_3/mmc$) is a layer of CoO_6 octahedra bound through the twin octahedra Sb_2O_9, as shown in the upper panel of Fig. 5.11 [220]. It is established that this compound is ordered antiferromagnetically below 3.8 K due to the interaction between layers [221]. The electron spin resonance method has established that this state is very non-trivial and is described within the framework of a six-sublattice antiferromagnet [219].

The integral of the exchange interaction was determined from the ratio $4.5J = g\mu_B B_S$ as $J = 18.2$ K, using $g = 3.8$. The plateau 1/3 on the magnetization curve lies in the field interval $0.306 < B/B_S < 0.479$ (the upper right panel in Figure 5.11), which corresponds to the theoretically predicted interval for the 2D antiferromagnetic triangular Heisenberg lattice in [211].

Three anomalies were observed at 3.82, 3.79, and 3.71 K in the temperature dependence of specific heat, shown in Fig. 5.12 [218]. Three magnetic phase transitions were associated with an easy-axis anisotropy, which leads to a sequence of transitions associated with

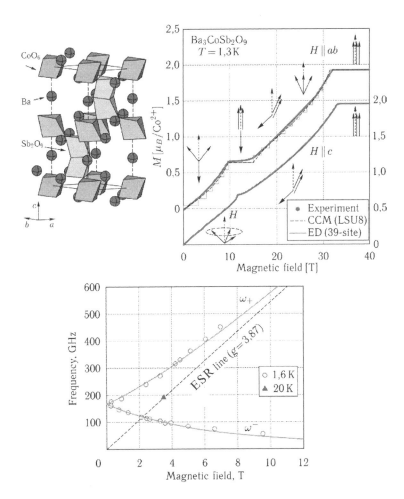

Fig. 5.11. The crystal structure (upper left) [220], the field dependence of the magnetization minus the van Vleck contribution (upper right) and the frequency–field diagram (bottom) for Ba$_3$CoSb$_2$O$_9$ [218, 219]. The solid lines on the right panel correspond to approximations within the framework of the six-sublattice antiferromagnet model.

the ordering of the z component, as shown in Fig. 5.13 in the form of a schematic phase diagram [220].

5.1.3. Delafossites

Delafossites are one of the most thoroughly investigated quasi-two-dimensional families, based on triangular geometry. These systems belong to a common class of layered compounds AMO_2, whose structure is the laying of MO_6 octahedra in MO_2 layers, between which monovalent cations A^+ are intercalated. Element M is a trivalent cation, whereas A is an alkali or noble(Ag, Cu, Pd, or Pt) [222]. Depending on the nature of this cation, its content (A_xMO_2, with $x \leq 1$), and synthetic procedures, sites of different intercalation of cations A between layers are observed: octahedral, trigonal-prismatic, tetrahedral, or linear (O–A–O dumbbell configuration).

A great interest of researchers in this group of materials is due to a number of exotic phenomena found in the delafossites. In particular, they exhibit multiferroelectric properties, as was found in the oxides $ACrO_2$ (A = Cu, Ag) [223–225], $AgFeO_2$ [226, 227], $CuFeO_2$ [228, 229], $CuFe_{1-x}M_xO_2$ (M = Al, Ga) [230–232] and $AgCrS_2$ [233, 234].

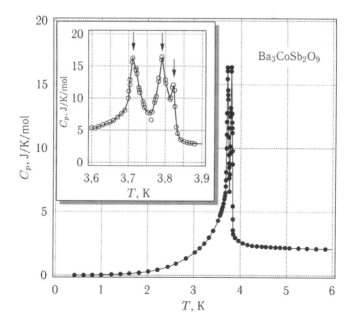

Fig. 5.12. The temperature dependence of specific heat $Ba_3CoSb_2O_9$ obtained in a zero field [218].

The magnetic structure in these compounds is extremely sensitive to the influence of an external magnetic field (Fig. 5.14). In the case of $CuMnO_2$, the anomalously strong dependence of the magnetic properties on the degree of doping was observed [235,236]. A remarkable phenomenon of a dimensional crossover from a 3D

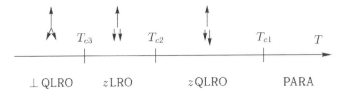

Fig. 5.13. Schematic magnetic phase diagram of a triangular antiferromagnet with an easy-axis anisotropy [220].

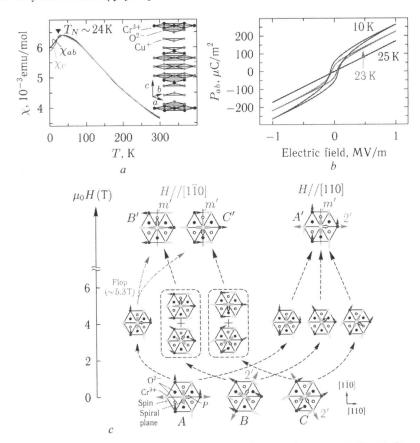

Fig. 5.14. Top panel: temperature dependence of magnetic susceptibility (left) and polarization loop $P(E)$ (right) [225]. The bottom panel: a schematic illustration of the evolution of the magnetic structure with variation of the external field for $CuCrO_2$ [224].

AFM to a 2D state of the spin-liquid type with low-energy magnetic excitations has been observed in the delafossite $Cu_{1-x}Ag_xCrO_2$ with increasing x [237].

5.2. BKT transition in two-dimensional magnetic materials with a triangular lattice

It is interesting to note that the transition of the BKT type on a triangular lattice was first discovered in the delafossites. The theory of the topological BKT transition applied to triangular 2D magnetic materials was developed much later than for 2D square systems in [238]. Calculations by the Monte Carlo method have shown that in triangular antiferromagnets topological defects (vortex–antivortex pairs) have a special nature, the so-called Z_2-vortices. Unlike the vortices in the BKT model with an arbitrary quantum number, that is, an arbitrary number of spin circulations inside the vortex, the Z_2-vortices are characterized by a topological quantum number that can only take two values. This type of vortex can be understood in terms of spin chirality as follows: in a 2D Heisenberg triangular antiferromagnet, frustration can be almost completely removed in a local 120° spin structure (Fig. 5.15). These short-range correlations give rise to an additional degree of freedom, characterized by a chirality vector, that is, by circulations clockwise or counterclockwise. As in the case of a 2D magnet with a square lattice considered above, in a triangular antiferromagnet at some finite temperature $T_{KM} \neq 0$ analogous to T_{KT}, we should expect a

Fig. 5.15. Schematic representation of Z_2-vortices in a triangular antiferromagnet.

topological transition associated with the formation of vortex–antivortex pairs Z_2. In this case, the temperature dependence of the correlation length follows the law:

$$\xi(T) = \xi_0 \exp\frac{b}{\tau^\nu}, \quad \text{where } \tau = \frac{T}{T_{KM}} - 1. \tag{5.4}$$

The calculated values of the parameter ν are $\nu = 0.42$ in the Kawamura and Miyashita model [238, 239] and $\nu = 0.37$ in the later model of Halperin et al. (KTHNY) [240], which is somewhat lower than $\nu = 0.5$ in the classical XY BKT model. The relationship between the transition temperature and the value of the exchange interaction is given in the triangular antiferromagnet by the relation

$$T_{KM} = 0.285 \left| \frac{2J}{k_B} \right| S^2. \tag{5.5}$$

An experimental confirmation of the possible realization of the BKT transition in triangular antiferromagnets was obtained by electron paramagnetic resonance. In particular, spin dynamics in accordance with the existence of Z_2-vortices was observed for $ACrO_2$ (A = Cu, Ag, Pd) (Fig. 5.16) [241], and also for $ACrO_2$ (A = H, Li, Na) [242]. The temperature of the BKT transition T_{KM} was determined from an analysis of the temperature dependences of the width of the ESR absorption line and was in good agreement with the relation (5.3) and the experimental values of the Néel temperature. It is interesting to note that for the 2D triangular antiferromagnet $LiCrO_2$ the anomalous peak at $0.9T_N \sim 56$ K was also observed in the muon relaxation rate (Fig. 5.17) [243]. This effect was also associated with a topological transition of the BKT type with the formation of Z_2-vortices.

The vortex phases predicted in the BKT theory are often called skyrmions in honour of the English physicist Skirme, who considered vortex structures to explain the properties of π-mesons. Skyrmions are one of the possible solutions describing equilibrium magnetic configurations in ferromagnetic and antiferromagnetic substances [244]. For the first time direct evidence of the formation of such vortex phases was detected by neutron diffraction in MnSi and $Fe_{12-x}Co_xSi$ when an external magnetic field was applied (Fig. 5.18a, b). The origin of skyrmions is similar to the formation of a lattice of cylindrical magnetic domains in classical micromagnetism: in the initial band structure (Figure 5.18 *a*), a region with a hexagonal packing arises (Fig. 5.18 *b*). The band structure corresponds to a magnetic helicoidal spiral, and the hexagonal structure corresponds to a vortex system.

The recent discovery of other skyrmion phases [247,248] attracted much attention to establishing the relationship between the complex magnetism in these compounds and their crystal symmetry, and in particular the connection between structural and magnetic chirality. In the above two archetypes: in MnSi and associated metallic silicides and germanides, the magnetic structure inherits the chirality of the non-centrosymmetric crystal structure through the antisymmetric Dzyaloshinsky–Moriya exchange, forming either a simple spiral (helicoid) or a more complex 3-*q* skyrmion phase complex in the applied magnetic fields. In contrast to the well-studied iron langasite $Ba_3NbFe_3Si_2O_{14}$ [249, 250], in these silicides the magnetic structure is helicoidal, but also possesses the chirality inherent in the crystalline structure of triangular 2D magnets. The magnetostructural coupling includes structural chirality, magnetic helicoidal and triangular magnetic chirality, which can be combined into a single phenomenological invariant.

5.2.1. Chiral and non-chiral polymorphs of manganese antimonate

The rare class of the family of skyrmion phases is also referred to as the manganese antimonate $MnSb_2O_6$ [251]. In the $MnSb_2O_6$ crystal structure, the layers of bounded distorted octahedra MnO_6 and SbO_6 form isolated triangular manganese plaques that alternate along the *c* axis with non-magnetic layers of SbO_6 octahedra, forming unfinished honeycomb planes of antimony ions (Fig. 5.19). $MnSb_2O_6$ resembles langasite [249, 250], and can be a multiferroic. Both polymorphs crystallize in the same space group *P*321, and have close structural units and ways of exchange interactions. However, the data on the magnetic structure of $MnSb_2O_6$ obtained by neutron diffraction differ from the magnetic structure of iron langasite and indicate that in $MnSb_2O_6$ the spin structure is based on cycloid rather than on helicoids, as shown in the right panel of Fig. 5.19.

The first-principles calculations in the theory of the density functional showed that the ground quantum state is determined by the family of supersuper exchange interactions between the nearest neighbors, and the spin-configuration model and the wave vector are

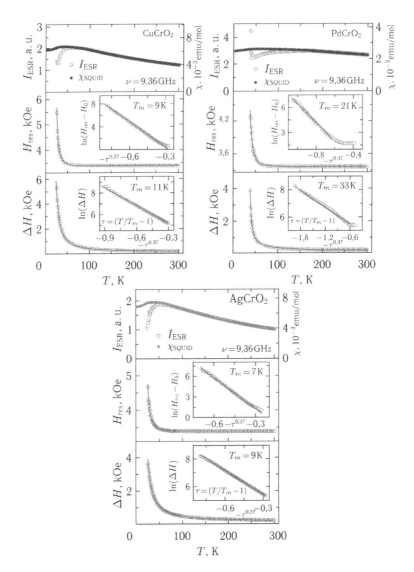

Fig. 5.16. Temperature dependences of the ESR parameters for $ACrO_2$ (A = Cu, Ag, Pd) [241]. Solid curves are the result of approximation in the framework of the BKT theory for triangular antiferromagnets.

in good agreement with the experimental neutron scattering data. In addition, it was suggested that $MnSb_2O_6$ may be a weakly polar multiferroic of an unusual form, since the change in the direction of the electric field should lead to a switching between one and two-domain configurations. The manganese ion Mn^{2+} is the only magnetic ion in the crystal, it has a spin of $S = 5/2$ and an orbital

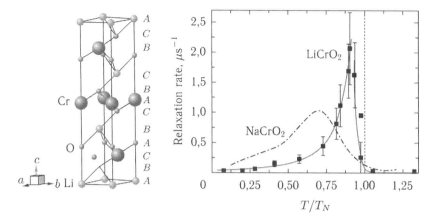

Fig. 5.17. The crystal structure (on the left) and the muon relaxation rate (on the right) of $LiCrO_2$ [243].

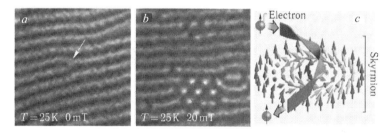

Fig. 5.18. Micrographs obtained by Lorentz transmission electron microscopy: strip helicoidal spin structure spontaneously formed in $Fe_{0.5}Co_{0.5}Si$ (arrow shows dislocation) (*a*); the appearance of a hexagonal lattice of vortices in an external field of 20 mT (the unit cell is separated by a hexagon) (*b*) [245]; effective 'topological' force acting on an electron moving through the skyrmion (*c*) [246].

angular momentum of $L = 0$. MnO_6 octahedra are isolated, and magnetic interactions occur with supersuper-exchange interactions involving two oxygen anions (Mn–O–O –Mn). The magnetization data indicate the presence of short-range magnetic correlations at temperatures below ~ 200 K, and the long-range antiferromagnetic order is established in the system at $T_N = 12.5$ K [252, 253]. Neutron diffraction studies have shown that in a magnetically ordered region, the magnetic moments of manganese form a 3D Heisenberg incommensurate spin structure with a wave vector $\mathbf{q} = (0.015, 0.015, 0.183)$ orthogonal to the *a* axis, which approximately corresponds to a cycloidal magnetic structure [251].

Manganese moments rotate in one plane containing the k_{exp}-cycloidal magnetic structure shown in the right panel of Fig. 5.19.

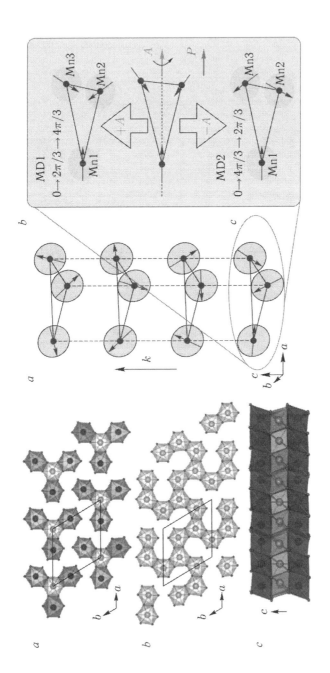

Fig. 5.19. Left panel: MnSb$_2$O$_6$ crystal structure: MnO$_6$ and SbO$_6$ octahedra are shown in different shades, oxygen ions are spheres [251]: *a*) a magnetoactive layer; *b*) an inactive layer of SbO$_6$; *c*) View along the *c* axis. Right panel: MnSb$_2$O$_6$ magnetic structure. Only one triangle of manganese ions is shown in the ab plane, a circle is indicated by a shade, described by spin rotation. *a*) Cycloids propagating along the *c* axis; *b*) and *c*) represent two magnetic domains, MD1 and MD2 respectively, generating axial rotation in a triangular coplanar mode.

The three cycloids in the unit cell have the same polarity, defined as $P_m = k \times (S \times S')$, where S and S' are the closest spins along the c axis. This situation should lead to ferroelectric polarization, by analogy with many other cycloidal magnets.

Subsequently, a new structural polymorph of manganese antimonate was synthesized – the metastable low-temperature phase of $MnSb_2O_6$ [254]. Differences in the structure of the crystal structure of the two polymorphs can be seen in Fig. 5.20. The structure of both polymorphs is layered with cationic ordering over the layers. In contrast to the chiral stable $MnSb_2O_6$ phase (space group $P321$) in the new non-chiral trigonal phase $MnSb_2O_6$ (space group $P31m$), all the antimony Sb^{5+} cations leave the magnetoactive layers, completing the 'honeycomb'. In this case, a two-dimensional layered structure is formed when layers of isolated manganese octahedra MnO_6 alternate with nonmagnetic layers of edge-bound SbO_6 octahedra forming planes with honeycomb order. Manganese ions in both compounds form an almost ideal triangular magnetic sublattice, so we can expect that in the presence of dominant antiferromagnetic exchange interactions the system will be frustrated.

An investigation of the magnetic susceptibility for both $MnSb_2O_6$ samples in weak magnetic fields $B = 0.1$ T revealed that the temperature dependence $\chi(T)$ exhibits an acute maximum with decreasing temperature, as shown in Fig. 5.21. This behavior indicates the appearance of a long-range antiferromagnetic order at low temperatures. Néel temperatures were determined from the maxima of $\chi(T)$ and were estimated as $T_N \sim 12$ K and ~ 8.5 K for the stable and metastable phases, respectively. In addition, on the $\chi(T)$ dependence the metastable phase exhibits a weak anomaly at a temperature of $T_1 \sim 41$ K, accompanied by the appearance of a small spontaneous magnetization. Approximation of the temperature dependence of the magnetic susceptibility in the temperature range 200–300 K using the Curie–Weiss law with allowance for the temperature-independent contribution gives the values of the Curie constant $C = 4.3$ emu K/mol and the Weiss temperature $\Theta \sim -20$ K for the stable phase and $C = 4.4$ emu K/mol and $\Theta \sim -17$ K for the metastable phase, respectively. The negative value of the Weiss temperature indicates the dominant antiferromagnetic interaction in these compounds. As the temperature is lowered, the $\chi(T)$ dependences deviate from the Curie–Weiss law, indicating an increase in the role of antiferromagnetic correlations as the temperature T_N is approached. The effective magnetic moment, estimated from the

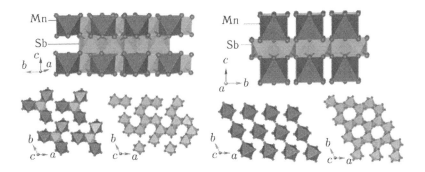

Fig. 5.20. Layered crystalline structure of $MnSb_2O_6$ stable phase ($P\,321$) (left) and metastable phase ($P31m$) (right): view along the c axis (upper line); projections of the magnetoactive and nonmagnetic antimony layer (bottom line). The octahedra MnO_6 and SbO_6 are shown in different shades.

Curie constant as $\mu_{eff} = 5.86\mu_B$/f.u. and $5.93\mu_B$/f.u. for stable and metastable phases, respectively, is in satisfactory agreement with theoretical estimates.

Investigations of the $\chi(T)$ dependences for the stable $MnSb_2O_6$ phase in external magnetic fields show a broadening typical of antiferromagnets and a weak shift of the anomaly at T_N toward lower temperatures. At the same time, detailed studies of the $\chi(T)$ dependences for the metastable phase of $MnSb_2O_6$ have shown that χ demonstrates a complex behaviour in the variation of the magnetic field (Fig. 5.22). In weak magnetic fields, the $\chi(T)$ dependence of the metastable phase of $MnSb_2O_6$ exhibits two anomalies at temperatures $T_N \sim 8.1$ K and $T_1 \sim 41$ K. With an increase in the magnetic field, the anomaly at T_1 is rapidly suppressed by an external field, but another anomaly arises at $T_C \sim 5$ K, which is observed up to about 1 T. Finally, in the field of strong fields, only the anomaly at T_N remains, which slightly shifts toward lower temperatures in a manner similar to the behaviour of $\chi(T)$ for the stable $MnSb_2O_6$ phase.

The behavior of the specific heat $C_p(T)$ in a zero magnetic field for the metastable phase of $MnSb_2O_6$ agrees with $\chi(T)$ in weak fields and shows two distinct anomalies, apparently connected with magnetic transitions. At a temperature $T_N = 8.5$ K, an anomaly of the λ-type is observed, which corresponds to a transition to an antiferromagnetic state. The nature of the second anomaly at T_1 is not completely clear. In magnetic fields, the position of the λ-type anomaly on $C_p(T)$ shifts toward lower temperatures (Fig. 5.23 a).

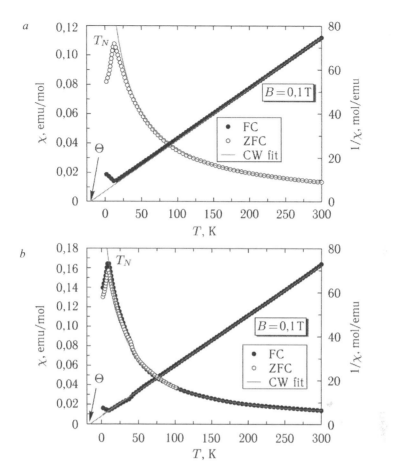

Fig. 5.21. Temperature dependences of the magnetic susceptibility and the reciprocal magnetic susceptibility $1/\chi$: *a*) of the stable phase of $MnSb_2O_6$, *b*) the metastable phase of $MnSb_2O_6$. Solid curves – approximation in accordance with the Curie–Weiss law

These data agree well with the maximum of the derivative of the magnetic susceptibility with respect to temperature.

The observed jump in the heat capacity during the phase transition is $\Delta C_p \approx 15.3$ J/mol K, which is noticeably lower than the expected value of 19.6 J/mol K from the mean-field theory and indicates the appreciable role of short-range magnetic correlations at temperatures above T_N, and frustrated systems. The jump in the magnetization ΔM at T_1 [255, 256] correlates with the anomaly in the heat capacity.

The magnetic contribution to the heat capacity C_m, shown in the inset of the bottom panel in Fig. 5.23 was obtained after subtracting the lattice contribution determined using an isostructural diamagnetic analog $ZnSb_2O_6$. Analysis of the $C_m(T)$ dependence below the Néel

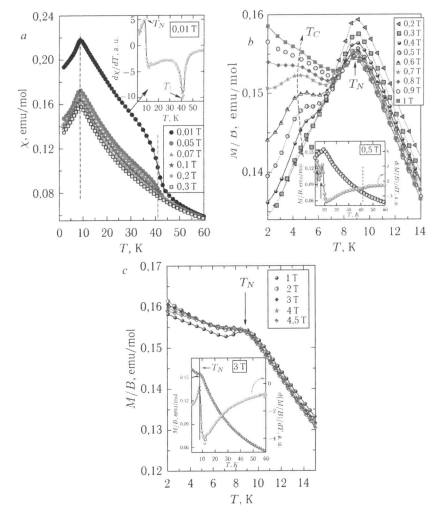

Fig. 5.22. The temperature dependence of the magnetic susceptibility for the metastable phase of MnSb$_2$O$_6$ upon variation of the magnetic field: *a*) weak field; *b*) the mean field; *c*) strong field.

temperature within the framework of the spin-wave theory confirms the presence of 3D AFM magnons at low temperatures. Magnetic entropy is saturated at a temperature of ~ 15 K at a level of 12 J/mol K, as shown in the inset on the bottom panel of Fig. 5.23. This value is slightly less than its estimate according to the mean-field theory of 14.9 J/mol K. A significant amount of entropy (more than 60%) is emitted significantly above the Néel temperature T_N, indicating a significant contribution of short-range correlations over a wide temperature range.

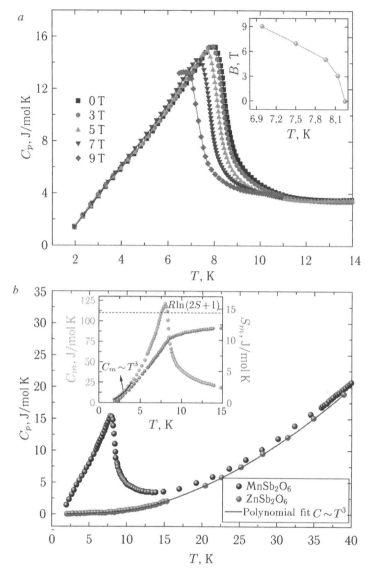

Fig. 5.23. The upper panel: the temperature dependence of the specific heat for the metastable phase of $MnSb_2O_6$ with the variation of the magnetic field in the low-temperature region. On the inset: the change in the position of the Néel temperature in a magnetic field to the maximum of the specific heat. The bottom panel: the specific heat at zero magnetic field for $MnSb_2O_6$ and $ZnSb_2O_6$. On the inset: the magnetic heat capacity and magnetic entropy at low temperatures and the approximation of the low-temperature part by the power law for magnons $C_m \sim T^{d/n}$.

When studying the ESR spectra on powder samples of two different modifications of $MnSb_2O_6$, it was found that in the whole investigated temperature range, a single exchange-narrowed Lorentzian line corresponding to the signal from Mn^{2+} ions is observed in the spectra. Careful analysis, however, showed that for the correct description of the shape of the metastable phase line at $T < 80$ K we should use the sum of two Lorentzians. This indicates the presence of an additional resonance mode, which manifests itself at low temperatures for a metastable sample and corresponds to the presence of a small amount of an impurity phase, as noted above.

The effective g-factor for the stable phase is practically independent of temperature and demonstrates an appreciable shift of the resonant field only in close proximity to T_N, while the value of the g-factor of the main mode of the metastable phase g_1 remains temperature-independent only up to $T \sim 50$ K, and is then deflected towards larger resonance fields. This indicates an increase in the role of short-range order correlations at temperatures exceeding the ordering temperature. The effective g-factor g_2 of the impurity mode exhibits pronounced anomalies and a deviation from $g = 2$ at $T < 80$ K.

An analysis of the temperature dependence of the line width $\Delta B(T)$ within the framework of the theory of critical broadening (2.15) gives a satisfactory description of the experimental data over a wide range of temperatures above the Néel temperature (15–300 K). The best agreement was obtained for the values of the model parameters presented in Table. 5.2. The values of the parameter T_N^{ESR} are close to the value T_N obtained from the measurement of static magnetic properties. The values of the critical exponent $\beta = 0.4$ and 0.5 for the stable and metastable phases, respectively, are in good agreement with $\beta = 1/3$ for the 3D Heisenberg antiferromagnet.

It was found that the magnetization of both $MnSb_2O_6$ samples reached saturation in moderate fields, and there was no hysteresis even at the lowest temperature reached in the experiment (Fig. 5.24). Magnetization curves $M(B)$, measured in pulsed fields at a temperature $T = 2.4$ K show saturation in the field of ~18.5 T for

Table 5.2 Néel temperatures and parameters obtained from an analysis of the temperature dependences of the ESR linewidth ΔB in accordance with Eq. (2.15)

Sample $MnSb_2O_6$	T_N (K)	T_N^{ESR} (K)	ΔB^* (mT)	β	A (mT)
Stable phase	12 ± 1	12 ± 1	7.0 ± 0.5	0.4 ± 0.1	7 ± 0.5
Metastable phase	8 ± 1	8 ± 1	12 ± 1	0.5 ± 0.1	5 ± 0.1

the metastable phase of $MnSb_2O_6$ and 26 T for the stable phase, and the saturation moment is close to the theoretically expected value for the high-spin manganese ion of $5\mu_B$. Analysis of the $M(B)$ dependences at low temperatures shows a noticeable change in the curvature in the fields ~ 1 T and 0.7 T at $T = 2$ K for the stable and metastable phases, respectively. This indicates the appearance of spin-reorientation transitions induced by the magnetic field. Moreover, two successive anomalies are seen on the derivative of the magnetization of the stable phase at $B_{C1} \sim 0.5$ T and $B_{C2} \sim 1$ T at $T = 2$ K (the left panel in Fig. 5.24). With increasing temperature, the position of B_{C1} remains almost unchanged, and the position of B_{C2} shifts toward larger fields. For a metastable phase, the critical field was $B_{SF} \sim 0.72$ T at $T = 2$ K.

The spin-configuration model for the stable phase of the antimonate $MnSb_2O_6$ was determined by the method of low-temperature neutron diffraction in [251]. It is established that the compound forms a unique chiral magnetic structure: in the magnetically ordered region, the magnetic moments of manganese form a 3D Heisenberg incommensurate spin structure with a wave vector $\mathbf{q} = (0.015, 0.015, 0.183)$ orthogonal to the a axis.

The basic exchange parameters and the spin-configuration model for the metastable phase of $MnSb_2O_6$ were obtained from first-principles calculations. It is interesting to note that this phase of the antimonate $MnSb_2O_6$ is an isostructural electronic analog of $MnAs_2O_6$ arsenate, for which neutron diffraction studies suggest long-range magnetic ordering below the temperature of ~ 12 K to an incommensurate spin structure with a wave vector $\mathbf{q} = (0,055, 0,389, 0,136)$ [257]. It should be noted that the magnetic properties of the metastable antimonate $MnSb_2O_6$ and arsenate $MnAs_2O_6$, although close, but not identical.

Figure 5.25 shows the supercell used for calculations of two neighbouring magnetoactive layers in the $MnSb_2O_6$ crystal structure and possible ways of exchange interactions between Mn^{2+} ions ($S = 5/2$). The shortest contact between the two nearest ions of manganese corresponds to supersuper exchange J_1 through two oxygen O...O as shown by arrows in Fig. 5.25a. J_2 denotes the integral of exchange interactions within the layer, while J_1 and J_3 are interlayer interactions. The values of the exchange integrals J_1–J_3 obtained are presented in Table 5.3.

All three exchange integrals J_1, J_2, J_3 are antiferromagnetic. J_1 is the strongest, but not dominant, since J_2 and J_3 are comparable to

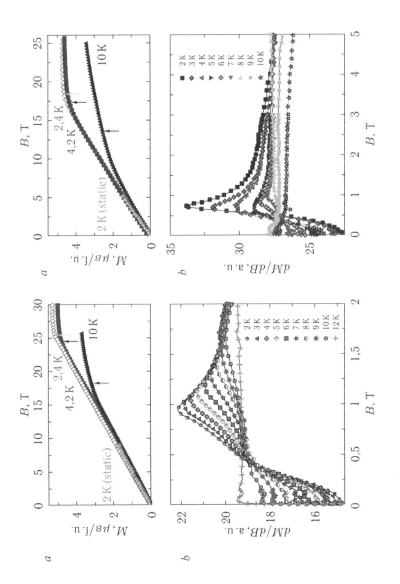

Fig. 5.24. The field dependences of the magnetization (*a*) and its first derivatives (*b*) measured in pulsed and static fields for the stable (left) and metastable (right) $MnSb_2O_6$ phases.

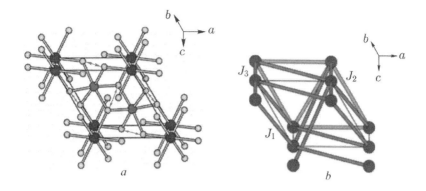

Fig. 5.25. A fragment of the crystal structure for two neighbouring magnetically active layers in $MnSb_2O_6$ (*a*) and the main possible paths of magnetic exchanges J_1, J_2 and J_3 between Mn^{2+} ions (*b*). Atoms Mn, Sb and O are shown by large, medium and small spheres, respectively.

Table 5.3. The exchange integral parameters and the Curie–Weiss temperature values (in Kelvins), obtained from the DFT + U calculations for $MnSb_2O_6$

	$U = 3$ eV	$U = 4$ eV	$U = 5$ eV
J_1/k_B	−1.28	−0.96	−0.72
J_2/k_B	−0.66	−0.54	−0.44
J_3/k_B	−0.72	−0.57	−0.46
Θ	−38.2	−29.4	−22.9

it in magnitude. Interestingly, the results suggest that the interlayer exchange interaction over the supersuper-exchange path is stronger than the intralayer, despite the greater distance between manganese cations. If the intralayer exchange integral J_2 is neglected, then the interlayer J_1 and J_3 form a two-dimensional antiferromagnetic structure. It should also be noted that all triangles (J_1, J_2, J_3) are frustrated. The results obtained are analogous to the data for $MnAs_2O_6$ [257]. Thus, it can be concluded that an incommensurate magnetic structure is realized in $MnSb_2O_6$, similarly to $MnAs_2O_6$.

Summarizing the data obtained in the study of the field and temperature dependences of the magnetic susceptibility, magnetization, and specific heat upon variation of the magnetic field, magnetic phase diagrams were constructed for the stable and metastable $MnSb_2O_6$ phases shown in Fig. 5.26. At temperatures of 12 K and 8 K in a zero magnetic field, a transition from the paramagnetic to the antiferromagnetic state for the stable and

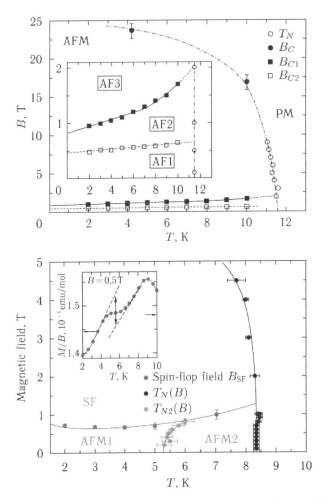

Fig. 5.26. The magnetic phase diagram for the stable (upper) and metastable (bottom) phases of $MnSb_2O_6$.

metastable phases of $MnSb_2O_6$, respectively. As the magnetic field increases, the phase boundary corresponding to this transition weakly shifts to low temperatures. In external magnetic fields ~ 1 T and ~ 0.5 T at $T = 2$ K for the stable and metastable phases, respectively, a spin-flop transition occurs to another antiferromagnetic phase, with two boundaries in the chiral phase: at B_{C1} and ~ 0.5 T and $B_{C2} \sim 1$ T. The saturation field of the magnetization is 18 T for the metastable phase and 26 T for the stable phase. The quantum ground state is antiferromagnetic, however, according to the results of theoretical calculations and neutron diffraction data, both phases are ordered into an incommensurate magnetic structure and are potential multiferroics. Both systems are sensitive to an external magnetic field, the application of which leads to the realization of various phases, apparently corresponding to different spin configurations.

5.2.2. Lithium–iron antimonate

The monoclinic crystal structure of the layered antimonate Li_4FeSbO_6, shown in Fig. 5.27, is a superstructure from a lattice of rock salt with cationic ordering over layers, the space group $C2/m$ [258].

In the structure of Li_4FeSbO_6, the magnetoactive mixed layers of iron, lithium and antimony cations in the octahedral oxygen environment alternate with non-magnetic layers of lithium, which is also in the octahedral environment of oxygen. Mössbauer spectroscopy showed that practically ideal cation ordering is established in the magnetoactive layer, so that around the SbO_6 octahedron there are three iron ions and three lithium ions, forming a hexagonal cell from the edge-shared alternating FeO_6 and LiO_6 octahedra (Fig. 5.27 *b*). At the same time, due to the replacement of half of the magnetic cations by non-magnetic Li, a triangular magnetic subsystem of Fe^{3+} ions is realized. There are no direct contacts of FeO_6 octahedra, and as the main exchange interaction a weak supersuper-exchange with the participation of non-magnetic cations Li^+ and/or Sb^{5+}.

The temperature dependence of the magnetic susceptibility in the field $B = 0.1$ T, presented in Fig. 5.28, shows a sharp maximum

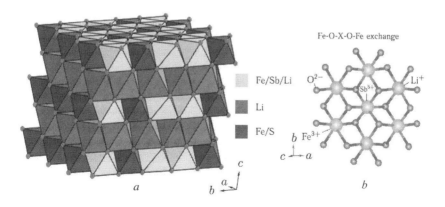

Fig. 5.27. The crystal structure of the layered antimonate Li_4FeSbO_6 – the polyhedral representation (*a*); fragment of the $C2/m$ structure in the *ab* plane, showing the triangular organization of Fe^{3+} cations in the magnetoactive layer of Li_4FeSbO_6 (*b*).

indicating the establishment of a long-range antiferromagnetic order at $T_N \sim 3.6$ K. The reciprocal magnetic susceptibility linearizes at temperatures above 100 K, which indicates the fulfillment of the Curie–Weiss law in the paramagnetic region. Negative and sufficiently large in absolute value (in comparison with the Néel temperature), the Weiss temperature $\Theta \sim -17$ K indicates a dominant antiferromagnetic interaction and a noticeable frustration ($f = \Theta/T_N \sim 5$) of the magnetic interaction. As the temperature is lowered, the dependence of $\chi(T)$ deviates from the Curie–Weiss law, thereby showing an increase in the role of antiferromagnetic correlations as the temperature T_N is approached. The obtained estimate of the effective magnetic moment $\mu_{eff} = 5.93\mu_B$/f.u. is in agreement with its theoretical estimate on the assumption that the magnetism in the compound Li_4FeSbO_6 is due to ions of iron Fe^{3+} in the high-spin state $S = 5/2$ with g-factor $g = 1.99$, determined from ESR data.

Investigations of the temperature dependences of the magnetization in the low-temperature region in magnetic fields up to 7 T have shown that the nature of the dependences becomes more complicated with increasing magnetic field strength (Fig. 5.29). In fields above ~ 0.5 T, the maximum on the temperature dependence of the magnetization is smeared out, and with a further increase in the field strength it splits into two maxima, the positions of which change substantially with the variation of the magnetic field. The more

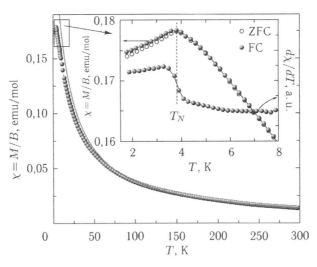

Fig. 5.28. Temperature dependence of the magnetic susceptibility of Li_4FeSbO_6. The solid line is the result of the Curie–Weiss approximation. On the inset: a low–temperature region.

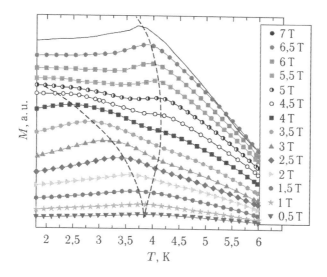

Fig. 5.29. Temperature dependence of the magnetization M for variation of the applied magnetic field for Li_4FeSbO_6. Dotted lines indicate the position of the anomalies and their displacement in external magnetic fields.

shallow and less intense left maximum (T_2) moves with increasing field toward low temperatures and is practically not detected in fields above 4.5 T in the temperature range 1.8–6 K, while the more acute right maximum (T_N) amplitude and shifts toward high temperatures to about 4.5 T, and with a further increase in the magnetic field shifts toward low temperatures. The presence of two features is most pronounced in intermediate fields $B \sim 3$–4 T, when the maxima diverge to the maximum distance. The magnetic-field-induced transition is probably associated with the reorientation of spins on a frustrated triangular lattice.

A study of the magnetization curves $M(B)$ showed that the magnetic moment in Li_4FeSbO_6 in fields up to 7 T is far from the saturation moment of the Fe^{3+} ion ($S = 5/2$) $M_{sat} \approx 5\mu_B$ (the upper panel of Fig. 5.30). At the same time, the analysis of experimental data reveals the presence of a wide maximum on the derivatives of the magnetization curves $dM/dB(B)$ (the bottom panel of Fig. 5.30), which corresponds to a change in the curvature of the field dependence in fields of about 3 T. The position of the maxima on the dependences $dM/dB(B)$, i.e., the critical field B_{SF}, corresponds to the position of the anomaly of T_2 on $M(T)$ (Fig. 5.29). With increasing temperature, the position of the maximum on the derivative of the magnetization over the field $dM/dB(B)$ in Li_4FeSbO_6 shifts toward

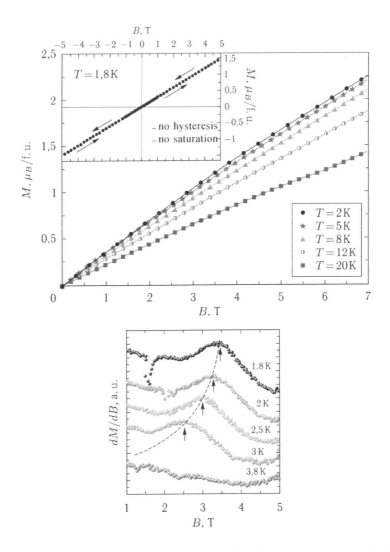

Fig. 5.30. Top panel: Magnetization curves for Li_4FeSbO_6 at temperature variation. On the inset: the complete $M(B)$ isotherm at 1.8 K. Bottom panel: field dependences $dM(B)/dB$ for the temperature variation. The dashed line and arrows show the position of the spin-flop transition detected at $T < T_N$.

smaller fields. In contrast to the sharp spin-flop transition observed in classical 3D antiferromagnets, the B_{SF} anomaly in Li_4FeSbO_6 is very broad. Similar anomalously wide spin-flop transitions were observed earlier in the $La_5Ca_8Cu_{24}O_{41}$ system [259].

To clarify the nature of the features found in the study of the temperature and field dependences of the magnetization in Li_4FeSbO_6 compounds, a study was made of the heat capacity in the temperature range 0.4–300 K when an external magnetic field was applied up to 9 T. Data on the specific heat in a zero magnetic field are in agreement with the temperature dependences of the magnetic susceptibility in weak fields and show an anomaly associated with the transition to an antiferromagnetic state at T_N. The character of this anomaly is different from the λ-type, and the additional feature (shoulder) differs at $T_m \sim 2.3$ K.

The magnetic contribution to the heat capacity C_m was estimated by subtracting the lattice contribution from the total heat capacity. Analysis of the low-temperature part of $C_m(T)$ in the framework of the theory of spin waves at $T < T_m$, according to which $C_m \sim T^{d/n}$, i.e., depends on the lattice dimension d and on the types of magnons n ($n = 1$ for antiferromagnets, $n = 2$ for ferromagnets). The best agreement is reached at $d \sim 2$ and $n = 1$, which indicates that the investigated compound is ordered into a 2D antiferromagnetic state. The magnetic entropy ΔS_m obtained by integrating the magnetic heat capacity saturates at a temperature of about 20 K, reaching ~ 15 J/mol K. This value is in agreement with the mean-field theory estimate of 14.9 J/mol K. Only $\sim 40\%$ of the entropy is released below the temperature T_N, which indicates a significant role of frustration and the contribution of short-range order correlations.

The results of studying the specific heat in external magnetic fields are shown in Fig. 5.31. As mentioned above, even in a zero magnetic field, two anomalies are observed at T_N and T_m. In weak magnetic fields, the anomaly at T_N is slightly shifted to higher temperatures. A further increase in the intensity of the external field leads to a change in this trend and a shift of T_N toward low temperatures. The position of the anomaly T_m is less sensitive to the variation of the external field.

The evolution of the ESR spectra by temperature for a powder sample of Li_4FeSbO_6 is shown in Fig. 5.32 at two different frequencies. It can be seen that the shape of the spectrum changes: the distortion and broadening of the ESR absorption line with decreasing temperature occurs below ~ 100 K as we approach T_N. Since the observed absorption line is relatively broad (only an order of magnitude smaller than the magnitude of the resonance field in a given compound), for accurate analysis line shape, it is necessary to take into account the presence of two components with circular polarization. The results of the approximation are in satisfactory agreement with the experimental data.

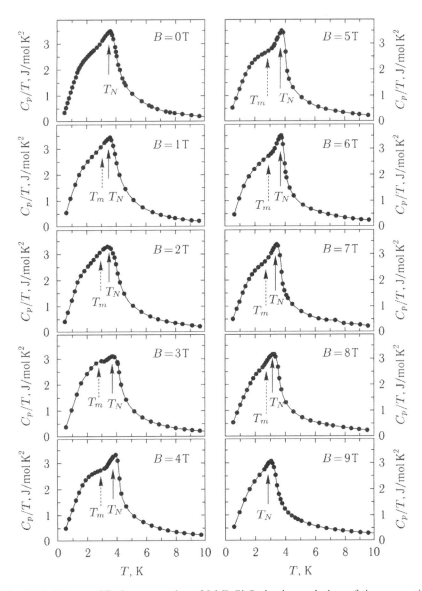

Fig. 5.31. The specific heat capacity of Li_4FeSbO_6 in the variation of the magnetic

At $T > 100$ K, the ESR spectra are satisfactorily described by a single Lorentzian, which corresponds to a signal from Fe^{3+} ions in an octahedral oxygen environment. The effect of the nearest neighbors of oxygen ligands on the ESR of the properties of Fe^{3+} ions can be represented by a spin Hamiltonian of the general form:

$$H = g\mu_B \vec{B} \cdot \vec{S} + b_2^0 \left[S_z^2 - \frac{1}{3} S(S+1) \right] +$$

$$\frac{1}{6} b_2^1 \left(S_x S_z + S_z S_x \right) + \frac{1}{3} b_2^2 \left(S_x^2 - S_y^2 \right),$$

(5.6)

where b_2^n is the splitting constant by the crystalline field, related to the symmetry of the ligand field. The parameters b_2^n can be expressed as a linear superposition of axially symmetric contributions from individual ligand ions. In octahedral complexes with symmetry O_h, the parameters b_2^n are close to zero ($b_0^2 = b_2^1 = b_2^2 = 1$), and an ESR absorption line near $g = 2$ should be expected [41]. Indeed, at high temperatures a single absorption line is observed, which is characterized by an isotropic g-factor $g = 1.99$.

The temperature dependences of the effective g-factor and linewidth, obtained from the approximation, and the frequency–field $f(B)$ diagram are shown in Fig. 5.33. The data obtained at different frequencies are consistent with each other. The dependence $f(B)$ at $T = 100$ K demonstrates the linear gapless behaviour typical for the paramagnetic state with the g-factor ~ 2. The effective g-factor remains practically constant up to low temperatures, and the visible shift of the resonance field, which is accompanied by degradation

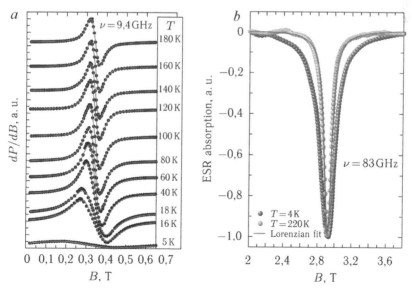

Fig. 5.32. The temperature evolution of the first derivative of the ESR absorption line (*a*) and the total absorption curve (*b*) for Li_4FeSbO_6 at two resonant frequencies.

of the absorption signal, appears only at helium temperatures. This behaviour indicates the opening of the gap for resonant excitations at the establishment of the long-range magnetic order.

At the same time, the most striking feature is the extended region of the substantial broadening of the resonant mode in practically the entire investigated temperature range. This behaviour indicates the presence of strong short-range order correlations substantially higher than T_N. The observed broadening of the absorption line is usually observed in antiferromagnets because of the decrease in spin fluctuations as the critical temperature is approached. In this case, the temperature changes ΔB can be described by the formula (2.15). The best agreement was reached with the following model parameters: $\Delta B^* = 10$ mT, $T_N^{ESR} = 3.56$ K and $\beta = 0.58$. The values of the parameter T_N^{ESR} close to the value of T_N obtained from the thermodynamic properties.

According to the theory of critical fluctuations, formula (2.15) must be valid up to temperatures $T \sim 3T_N$. However, in the investigated compound this region is more extended. A similar result was observed earlier for other strongly correlated systems. In the framework of the Dormann and Jaccarino approach [102], the temperature behaviour of ΔB in the interval $3T_N < T < 10T_N$ can be described using the formula:

$$\Delta B(T) = \Delta B_0 \cdot \left[\chi_0(T) / \chi(T) \right],$$ (5.7)

where $\chi_0 = C/T$ is the susceptibility of the free ion, C is the Curie constant, and $\chi(T)$ is the statistical susceptibility of the interacting system. Using the obtained data on the static magnetic susceptibility and the Curie constant, estimated as 4.3 emu K/mol, and also the value of $\Delta B_0 = 38.6$ mT in the high-temperature limit, the theoretical dependence of $\Delta B_0 \ [\chi_0(T)/\chi(T)]$ was compared with the experimental data (dashed curves in Fig. 5.33). It can be seen that in the low-temperature region this approximation substantially deviates from the experimental data.

For an alternative analysis of the broadening of the absorption line, it is possible to describe the critical behaviour of $\Delta B(T)$ in the context of the BKT scenario. The width of the ESR absorption line within the framework of the BKT model can be expressed as

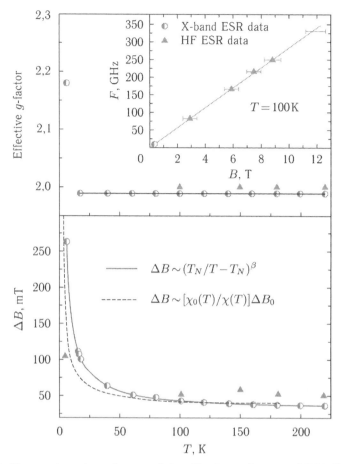

Fig. 5.33. Temperature dependences of the effective g-factor and ESR linewidth ΔB for Li_4FeSbO_6. In the inset $F(B)$, the diagram at 100 K. The solid and dashed curves at the bottom of the figure are the result of approximating the dependences of $\Delta B(T)$ in the framework of the model of critical fluctuations.

$$\Delta B = \Delta B_\infty \exp\left(\frac{3b}{\tau^\nu}\right), \qquad (5.8)$$

where ΔB_∞ is the asymptotic high-temperature value of the line width, b is an arbitrary parameter, and the values of the parameter ν in the case of a triangular spin lattice are $\nu = 0.42$ in the Kawamura and Miyashita model [238, 239] and $\nu = 0.37$ in the model of Halperin et al. [240]; τ denotes the normalized temperature $\tau = T/T_{KM} - 1$, where T_{KM} is the temperature of the topological phase transition.

The results of the approximation in the framework of the Kawamura–Miyashita approach are summarized in Table 5.4. The

Table 5.4. The parameters obtained from the approximation of the $\Delta B(T)$ dependence within the framework of the Kawamura and Miyashita model

ν	ΔB_∞, mT	T_{KM}, K	b	R^2
0.37	18.7	1.0	1.7	0.9999
0.42	22.8	1.1	1.6	0.9992

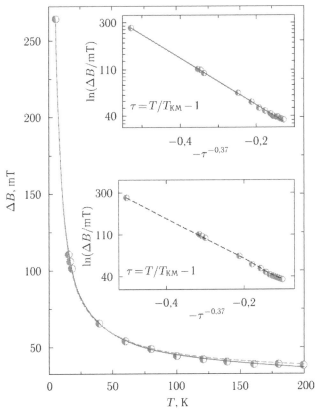

Fig. 5.34. Temperature dependence of the width of the ESR line for Li_4FeSbO_6 Solid and dashed curves are approximations of the $\Delta B(T)$ dependence within the framework of the Kawamura and Miyashita model. Inserts: The logarithmic dependences of the ESR linewidth $\ln(\Delta B)$ on the normalized temperature $-\tau^\nu$.

behaviour of the line width is well described for both exponentials ν in the whole investigated temperature range (Fig. 5.34), but the value $\nu = 0.37$ is preferable, starting from the convergence parameter R^2. As expected, the T_{KM} temperature assumes a value below T_N. The parameter b was close to the theoretically predicted value for the

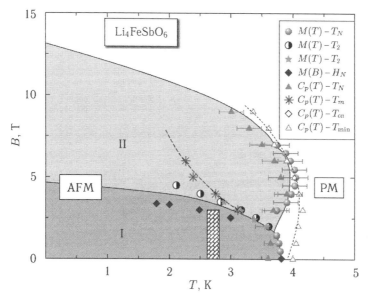

Fig. 5.35. Magnetic phase diagram of the layered antimonate Li_4FeSbO_6.

BKT regime $b = \pi/2$ and is in agreement with the values $b = 1.86$ and 1.93 obtained for 2D triangular antiferromagnets $CuCrO_2$ and $AgCrO_2$ [241].

Summarizing the data of thermodynamic and resonance studies, we can construct a magnetic phase diagram for the new layered antimonate Li_4FeSbO_6, shown in Fig. 5.35. The character of the motion of the AFM–PM phase boundary indicates the low-dimensional nature of the magnetic correlations and the appreciable frustration in Li_4FeSbO_6, when the external magnetic field acts as an additional ordering factor on the spin subsystem and raises the temperature of magnetic ordering due to frustration suppression. The quantum ground state for the investigated compound is the antiferromagnetic state of AFM I, however, in magnetic fields it is

Quasi-two dimensional magnets with a honeycomb magnetic lattice

6.1. Frustration due to the competition of exchange interactions

A large number of the currently known layered metal oxides are characterized by the geometry of cation ordering in magnetically active layers of the honeycomb-type. This structural type is a variant of the organization of a triangular magnetic lattice, but geometric frustration in this case is removed, since the number of spins in a hexagonal cell is doubled in comparison with triangular. At the same time, it is important to note that the causes of frustration can be not only geometric but also have an exchange nature. In this case, frustration arises from the competition of exchange interactions, when the interaction becomes important not only between the nearest neighbours, but also the nearest neighbours or third, etc. It is well known that in the case of the classical Heisenberg model ($S = \infty$) on a honcycomb-type lattice, taking into account the exchange interaction only between the nearest neighbors J_1, the Néel antiferromagnetic quantum ground state is realized [260] (Fig. 6.1). In the case, however, when the exchange interactions with the next neighbours (second J_2 and/or third J_3) play an important role and have an antiferromagnetic nature, frustration arises in the magnetic system, which substantially complicates the general picture of the ground state. Depending on the sign and the magnitude of the J_2/J_1 and J_3/J_1 ratios, non-trivial spin configurations, such as zigzag

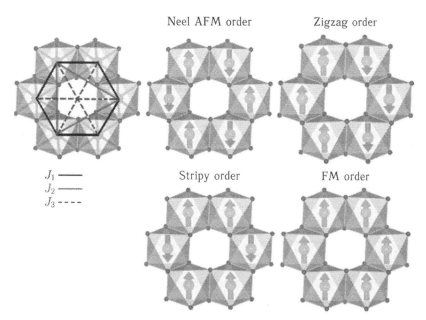

Fig. 6.1. 2D honeycomb lattice in the Heisenberg model with exchange interactions up to the third order, $J_{1,2,3}$, and the simplest spin-configuration diagrams.

ordering, stripe, various spiral structures, etc., can occur [261].

The classical magnetic phase diagram calculated for 2D Heisenberg magnets on a honeycomb-type lattice, depending on the relations J_2/J_1 and J_3/J_1 in the model J_1–J_2–J_3 (for $J_1 = 1$, $J_2 > 0$ and $J_3 > 0$) and the corresponding spin models are shown schematically in Fig. 6.2 [262]. According to the calculations of [262], for $J_3 > 0.6$ there is a first-order transition between the ground state of the Néel AFM type and the AFM zigzag state. Spiral phases stabilize over a wide range of J_2 values. In addition, numerical calculations predict a phase with no magnetic order near $J_2 = J_3 = 0.5$. The nature of the ground state remains open in this case.

According to the authors of [262], the stabilization of the dimerized state is unlikely. At the same time, calculations by the Monte Carlo method in the J_1–J_2 model in the Hubbard approximation (taking into account the large Coulomb repulsion with the potential U, i.e., the case of the Mott insulator) reveal a wide range of spin-liquid and dimerized phases for 2D magnetic materials with a honeycomb lattice [263]. The obtained estimates for the critical ratios of the exchange parameters for the phase boundaries between the Néel and spin-liquid states and between the spin-liquid and

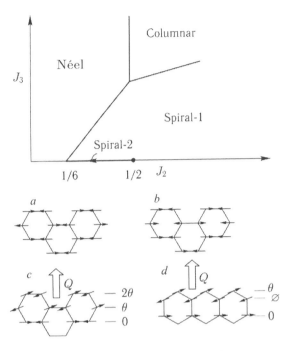

Fig. 6.2. The magnetic phase diagram for 2D Heisenberg magnets on the bee honeycomb lattice in model J_1–J_2–J_3 and the corresponding spin models: *a*) Néel's AFM (Néel), *b*) AFM zigzag (columnar), *c*) and *d*) spiral-1) and (spiral-2) [262].

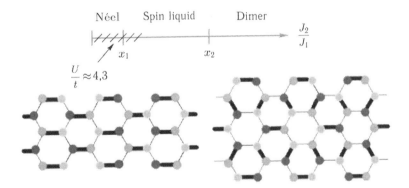

Fig. 6.3. The phase diagram for 2D Heisenberg magnets on the bee honeycomb lattice in model J_1–J_2 (top) [263] and dimerized spin models (bottom) [264].

the dimerized states were $J_2/J_1 = x_1 = 0.08$ and $J_2/J_1 = x_1 = 0.3$, respectively (the lower part of Fig. 6.3).

6.2. Kitaev model

An important role among the quantum theoretical models for 2D magnets with a honeycomb lattice is played by the Kitaev model, in which the exact solution for the honeycomb lattice of spins $S = 1/2$ [265] is obtained. Kitaev considered the problem, assuming that there are three types (x, y and z) of the paths of exchange interactions in the cell of the honeycomb as shown on the left side of Fig. 6.4. Then the Hamiltonian with exchange parameters J_x, J_y and J_z is written as

$$H = -J_x \sum_{x-\text{links}} \sigma_j^x \sigma_k^x - J_y \sum_{y-\text{links}} \sigma_j^y \sigma_k^y - J_z \sum_{z-\text{links}} \sigma_j^z \sigma_k^z. \tag{6.1}$$

The ratios of the exchange parameters J_x, J_y and J_z, regardless of their sign, determine the region of solutions of the eigenvalues of the energies for the indicated on the right-hand side of Fig. 6.4 of the unit cell, in the form of the phase diagram shown in Fig. 6.5. Region B on the phase diagram, shown by a white triangle, is determined by the inequalities:

$$|J_x| \leqslant |J_y| + |J_z|, \quad |J_y| \leqslant |J_x| + |J_z|, \quad |J_z| \leqslant |J_x| + |J_y|,$$

and corresponds to a gapless state of the spin-liquid type. The regions A_x, A_y, and A_z, shown by shaded triangles, correspond to the gap quantum ground states. According to the Kitaev model, in the presence of anisotropy on the lattice, bee honeycombs of spins $S = 1/2$ can realize four possible types of ground state: FM, Néel AFM,

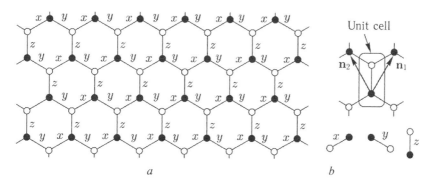

Fig. 6.4. The Kitaev model: three types of connections in the honeycomb lattice (a) and the unit cell used for the calculation, where $\mathbf{n}_1 = (1/2, \sqrt{3}/2)$ and $\mathbf{n}_2 = (-1/2, \sqrt{3}/2)$ in standard xy-coordinates (b) [265].

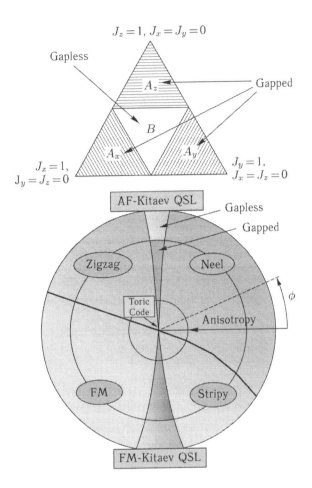

Fig. 6.5. The upper panel: the phase diagram of the Kitaev model. The triangle corresponds to the section of the positive octant (J_x, J_y, $J_z \geq 0$) by the plane $J_x + J_y + J_z = 1$. Diagrams for other octants are similar [265]. Bottom panel: phase diagram of the Kitaev–Heisenberg model with allowance for the anisotropy [266].

AFM zigzag and AFM stripe, as shown in the bottom panel of Fig. 6.5 [266].

6.3. Experimental realization of Kitaev's model

The most popular candidates for checking the Kitaev model now are metal oxide compounds with $5d$ elements, primarily sodium and lithium iridates Na_2IrO_3 and Li_2IrO_3. It was theoretically predicted [267] that for a strong spin-orbit interaction, these systems can be an example of a new state that implements the Kitaev model. In calculations [267], the Hamiltonian of the problem (the Kitaev–Heisenberg model) included two terms: corresponding to anisotropic interactions (the analogue of the Kitaev Hamiltonian (6.1)) and the Heisenberg term corresponding to isotropic interactions between nearest neighbours. When writing through a single parameter α, the Kitaev–Heisenberg Hamiltonian can be written in the form:

$$H_{ij}^{(\gamma)} = -2(2\alpha - 1)\tilde{S}_i^{\gamma}\ \tilde{S}_j^{\gamma} - (1 - \alpha)\tilde{S}_i \cdot \tilde{S}_j, \qquad (6.2)$$

where $J_1 = 2\alpha$, $J_2 = 1 - \alpha$. The situation when $\alpha = 0$ corresponds to a purely Heisenberg limit, and $\alpha = 1$ – to the purely Kitaev limit; $\alpha = 1/2$ corresponds to the ferromagnetic state. Numerical calculations for all values of $0 \leq \alpha \leq 1$ allow us to construct the phase diagram of the quantum ground states shown in the top panel of Fig. 6.6 The magnetic field effect was taken into account in [268].The resulting phase diagram is shown in the bottom panel of Fig. 6.6.

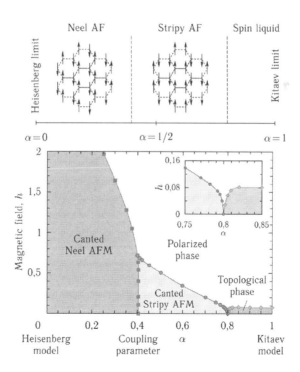

Fig. 6.6. Magnetic phase diagram of the ground quantum state of a 2D magnetic material with a honeycomb lattice, calculated in the Kitaev–Heisenberg model [267] (top panel); phase diagram taking into account the influence of an external magnetic field [268] (bottom panel).

The crystal structure of both Na_2IrO_3 and Li_2IrO_3 oxides is layered (monoclinic symmetry, space group $C2/m$), with the magnetic layers alternating with non-magnetic layers of sodium. Iridium ions Ir^{4+} with spin $S = 1/2$ in the octahedral environment form an almost ideal lattice of honeycomb cells, shown in Fig. 6.7 [269]. An investigation

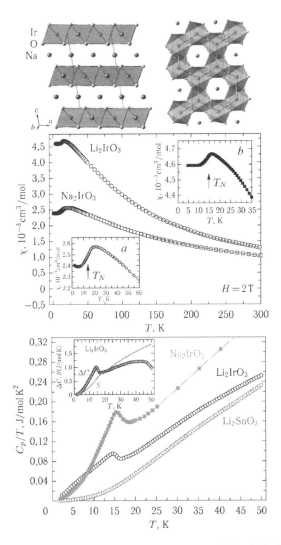

Fig. 6.7. Top panel: layered crystalline structure of Na_2IrO_3 [269]. The middle panel: the temperature dependences of the magnetic susceptibility, on the insets – the maximum shape on an enlarged scale for Na_2IrO_3 (*a*) and Li_2IrO_3 (*b*). The lower panel: the temperature dependences of specific heat capacity of Na_2IrO_3, Li_2IrO_3, and the diamagnetic structural analogue Li_2SnO_3; on the inset – the calculated magnetic contribution to specific heat capacity and magnetic entropy in Li_2IrO_3 [270].

of the thermodynamic properties indicates the establishment of a long-range AFM of the order at the same Néel temperature $T_N \sim 15$ K for both compounds, but with significantly different Weiss temperatures (-125 K for Na_2IrO_3 and -33 K for Li_2IrO_3) [270]. In this case, the phase transition to the ordered state precedes a broad maximum on the temperature dependence of the magnetic susceptibility, characteristic of quasi-2D magnetic systems (Fig. 6.7).

Despite the predicted possible spin-liquid state, direct investigations by neutron and magnetic resonance x-ray diffraction have shown that the ground state for sodium iridate Na_2IrO_3 is an antiferromagnet of the zigzag type [271], which contradicts the orderings of the Néel and stripe-type expected from the Kitaev–Heisenberg model 6.2), and an incommensurable non-coplanar spiral antiferromagnet for Li_2IrO_3 [272]. It is established that, in spite of the almost ideal lattice of the honeycomb-type, IrO_6 octahedra are tested in Na_2IrO_3, a strong trigonal distortion, which leads to a noticeable deviation of the angle of Ir–O–Ir bonds from the ideal angle 90° for the dominance of the Kitaev exchange (Fig. 6.8 *a*). This circumstance plays a key role in stabilizing the state of the antiferromagnetic zigzag type in this compound. At the same time, in the lithium analogue, the trigonal distortions of iridium octahedra

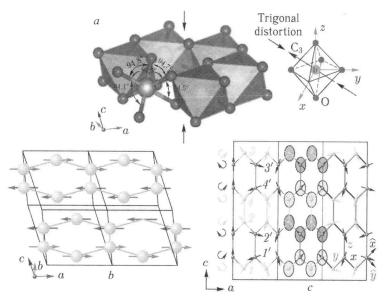

Fig. 6.8. Local structure and trigonal distortion of octahedra in the compound Na_2IrO_3 (*a*). Spin-configuration model: AFM zigzag in the Na_2IrO_3 compound (*b*) [271]. The spiral spin structure in Li_2IrO_3 (*c*) [272].

are much smaller, so it is closer to the spin-liquid regime predicted in the Kitaev–Heisenberg model [270] (Fig. 6.8 *c*).

A candidate for testing the Kitaev model was also a compound with 4*d* ions α-RuCl$_3$ ion (monoclinic symmetry, *C2/m* space group) in which the ruthenium ion is present in the low-spin state (*S* = 1/2). The edge-shared RuCl$_6$ octahedra form in this compound an almost ideal lattice of the honeycomb-type, shown in Fig. 6.9 *a*, and the layers are weakly coupled by the van der Waals forces [273]. Detailed studies of the thermodynamic properties (Figures

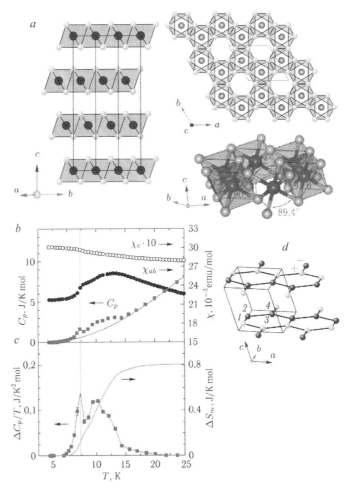

Fig. 6.9. The layered crystal structure of RuCl$_3$ [273] (*a*); temperature dependences of the specific heat and magnetic susceptibility for two field directions: along the *c* axis and in the *ab* plane [275] (*b*); the magnetic part of specific heat (points) and the magnetic entropy (line) released during the phase transition (*c*); spin model – AFM zigzag for RuCl$_3$ [274] (*d*).

6.9*b*, *c*) and experiments on neutron scattering by single-crystal samples indicate the establishment of a long-range antiferromagnetic order of the zigzag type at a temperature $T \sim 13$ K (Fig. 6.9 *d*) [274] in agreement with the Kitaev model which assumes this type of ordering to be one of the possible quantum ground states. This connection is also interesting by the appearance of a spin-reorientation metamagnetic transition induced by a magnetic field at $B \sim 8$ T (Fig. 6.10).

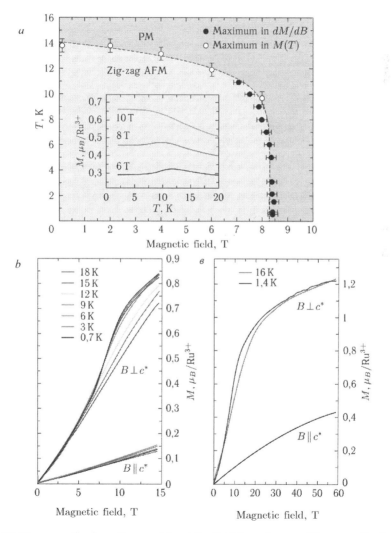

Fig. 6.10. The magnetic phase diagram (*a*) and the field dependences of the magnetization (*b*, *c*) in RuCl$_3$ [274].

Another quasi-two dimensional magnet with ruthenium, $SrRu_2O_6$ (hexagonal symmetry, space group $P\bar{3}1/m$), based on the geometry of honeycomb cells, is characterized by an unusually high Néel temperature, $T_N \sim 560$ K, for the layered system [276]. In addition, it was found that the experimentally observed magnetic moment of $1.3\mu_B$/Ru is half that of the Ru^{5+} ion ($4d^3$, $S = 3/2$). To explain this behaviour, theoretical models were proposed in which the system was considered as a delocalized magnet, and the high T_N was explained by the presence of interplanar exchange interaction or single ion magnetic anisotropy, respectively [276, 277]. A detailed study of the electronic structure showed that the anomalous magnetic properties are most likely associated with the formation of Ru_6-quasimolecular orbitals (molecular orbitals of different symmetries formed on six Ru atoms) on an ideal lattice of honeycomb-type in $SrRu_2O_6$. This leads to a substantially non-Heisenberg type of the exchange Hamiltonian. The simplest estimates showed that the interplanar exchange interaction exerts the strongest influence on T_N. Also for this compound was found a strong dependence of the magnetic properties of the system on doping and predicted the appearance of a semi-metallic state (the gap opens only in one spin subband).

A very unusual quantum ground state is formed in a structurally related Li_2RuO_3 oxide described above, in which Ru^{4+} ions ($4d^2$, $S = 1$) form honeycomb layers separated by lithium layers [278]. As the temperature is lowered to ~ 270 K, Li_2RuO_3 undergoes a structural phase transition between two monoclinic modifications (from the space group $C2/m$ to $P2_1/m$), accompanied by a jump in the volume of the unit cell, as shown in the upper panel of Fig. 6.11.

Experimental studies of the structural and magnetic properties, as well as theoretical calculations for this low-dimensional magnet, have shown that in the high-temperature phase ($T > 270$ K) in Li_2RuO_3, a state of the valence bond liquid is realized [278]. Local dimers Ru–Ru form in this dimerized state but they are, however, dynamically disordered in the honeycomb-type lattice. With a decrease in temperature, the dimers 'freeze' in certain positions (the lower panel of Fig. 6.11 *a*, *b*). In the dimerized low-temperature phase, Li_2RuO_3 is a non-magnetic system (spin-singlet state in dimers from spins $S = 1$). With an increase in temperature, a transition to the paramagnetic state is observed, which corresponds to the existence of isolated spins $S = 1/2$, rather than spins $S = 1$, characteristic for the Ru^{4+} ions. This behaviour is attributed to the stabilization of the orbital-selective state in which some electrons participate in the

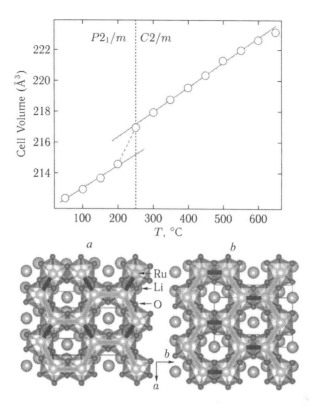

Fig. 6.11. The temperature dependence of the unit cell volume from neutron diffraction data (top) and the theoretically calculated dimerized spin states in Li_2RuO_3 (bottom) [145].

formation of the spin-singlet state, while the other part behaves like local spins.

A good example of a non-magnetic two-dimensional behaviour is bismuth manganese oxynitrate $Bi_3Mn_4O_{12}(NO_3)$ [279, 280]. The compound crystallizes in the trigonal space group $P3$ and is characterized by a layered crystal structure in which the layers of the manganese ions Mn^{4+} ($S = 3/2$) are organized in a honeycomb-type topology, as shown in the top panel of Fig. 6.12.

The temperature dependence of the magnetic susceptibility, shown in the right lower panel of Fig. 6.12, indicates a wide correlation maximum at $T \sim 70$ K, which is characteristic of the 2D AFM systems. Weiss temperature assumes a large negative value of $\Theta \sim -257$ K, however, the compound does not exhibit a long-range magnetic order up to 0.4 K, which was attributed by the authors [279] to strong frustration because of the competition between J_1 (between the nearest neighbours) and J_2 between the next-neighbours). Experiments on neutron scattering in magnetic fields revealed an ordering induced under a magnetic field in the AFM phase of the Néel type, which is realized at $B \sim 5$ T [280]. The magnetic structure induced by the external field is shown in the right panel of Fig. 6.12.

6.4. BKT transition in the honeycomb lattice

Similar to the 2D magnetic materials described above with square and triangular lattices, in magnetic materials based on the honeycomb topology, indications were found for realizing the topological transition of the BKT. Thus, an unusual spin dynamics for a 2D antiferromagnet $BaNi_2V_2O_8$ with a honeycomb-type lattice on nickel

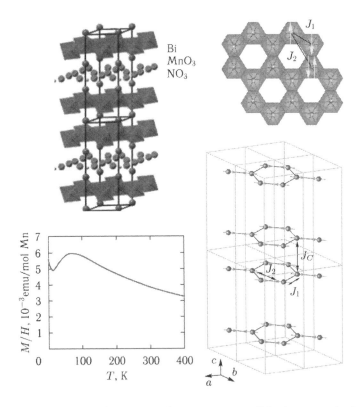

Fig. 6.12. A layered crystal structure, the arrangement of the honeycomb plane, the temperature dependence of the magnetic susceptibility, and the exchange interaction paths J_1, J_2, J_C in $Bi_3Mn_4O_{12}(NO_3)$ [279]. Arrows show the structure of the Néel type, which is established under the action of a magnetic field [280].

ions ($S = 1$) has been established from the study of the temperature dependences of the width of the ESR absorption line, which is proportional to the cube of the spin-correlation length [281] (Fig. 6.13). The compound has trigonal symmetry, space group R-3. The experimental results were interpreted within the framework of the BKT model and a conclusion was made that the investigated object is a weakly anisotropic easy-plane 2D Heisenberg antiferromagnet. The temperature of the BKT transition and the critical exponent were

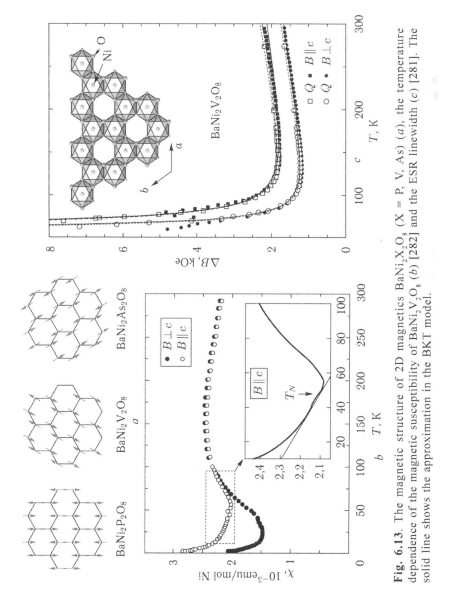

Fig. 6.13. The magnetic structure of 2D magnetics $BaNi_2X_2O_8$ (X = P, V, As) (a), the temperature dependence of the magnetic susceptibility of $BaNi_2V_2O_8$ (b) [282] and the ESR linewidth (c) [281]. The solid line shows the approximation in the BKT model.

$T_{KT} \sim 43$ K and $\beta = 0.9$, respectively, which agrees well with the presence of a broad 2D correlation maximum at $T_{max} \sim 125$ K and 3D long-range AFM ordering at $T_N = 50$ K on the temperature dependences of the magnetic susceptibility [282] .

6.5. Antimonates and tellurates of transition metals

Among complex quasi-two dimensional oxides, tellurates and antimonates of alkali and transition metals with a layered structure in which ordered mixed layers of magnetic cations and antimony (or tellurium) alternate with layers of alkali metal cations are of great interest [283–314]. In this case, magnetic cations are in most cases ordered in two basic structural variants. For the first type, the cations M form layers of bounded octahedra in the compounds $A_3^+M_2^{2+}X^{5+}O_6$ and $A^{2+}M_2^{2+}Te^{6+}O_6$ (A = Li, Na, Ag; X = Bi, Sb; M –transition metal) (honeycomb-type lattice), in which 1/3 positions are occupied by non-magnetic X ions (for example, tellurium or antimony). At the same time, for nonmagnetic lithium or sodium cations alternating with transition metal cations A_4MXO_6(A = Li, Na; X = Sb, Te; M = Co, Ni, Fe), and then a variant of the triangular arrangement of the magnetic cations is realized. Both these structural types are the variants of the organization of a triangular magnetic grating.

The tellurates and antimonates exhibit a wide variety of magnetic properties, including magnetic ordering at low temperatures [295–302], spin-gap and spin-glass behaviour [303–312], spin-reorientation and metamagnetic transitions [295, 300, 301].

Among the 2D compounds studied in this family with a honeycomb-type lattice, the long-range magnetic order at low temperatures, was detected in the antimonates $Na_3M_2SbO_6$ (M = Co, Ni) [295, 296], bismuthates $A_3Ni_2BiO_6$ (A = Li, Na) [299, 301], tellurates $Na_2M_2TeO_6$ (M = Co, Ni) [295 , 297, 310], as well as in systems with the delafossite-type structure $Cu_3M_2SbO_6$ (M = Co, Ni) [300, 302].

At low temperatures the compounds of $Na_3M_2SbO_6$ (M = Co, Ni), $Na_2M_2TeO_6$ (M = Co, Ni) and $A_3Ni_2BiO_6$ (A = Li, Na) (monoclinic symmetry, space group $C2/m$) exhibit characteristic behaviour for uniaxial antiferromagnets (Fig. 6.14). Doping leads to a significant change in the magnetic properties in these compounds and the suppression of the long-range magnetic order. In particular, polycrystalline samples of $Li_3NiMBiO_6$ (M = Mg, Cu, Zn) exhibit paramagnetic behaviour over the entire temperature range, whereas

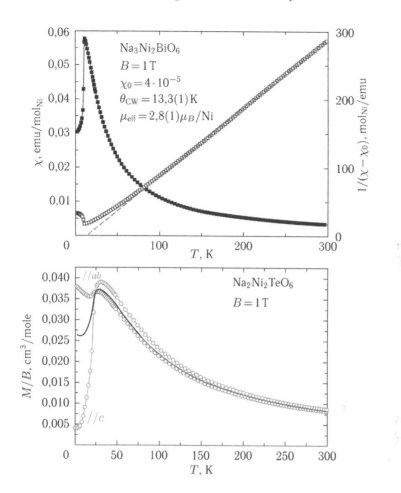

Fig. 6.14. Temperature dependences of the magnetic susceptibility for a polycrystalline $Na_3Ni_2BiO_6$ sample (upper panel) [301] and a single crystal $Na_2Ni_2TeO_6$ sample (lower panel) [310].

the original $Li_3Ni_2BiO_6$ bismuthate shows an AFM ordering with $T_N \sim 5.5$ K with an effective moment of $4.52\mu_B$ per formula unit (Fig. 6.15) [299]. For $Li_3NiMgBiO_6$ and $Li_3NiZnBiO_6$ substitution of half of nickel ions for nonmagnetic cations lowers the value of the effective magnetic moment to 2.84 and $2.90\mu_B$, respectively. In $Li_3NiCuBiO_6$, the experimental value of the total effective moment of $3.38\mu_B$ agrees satisfactorily with the theoretical value for the total action of Ni^{2+} and Cu^{2+} cations taking into account the spin-only magnetic moment ($\mu_{theor} = 3.32\mu_B$).

In [297], detailed studies of the static magnetic properties of solid solutions of $Na_2Ni_{2-x}Zn_xTeO_6$, $Na_2Ni_{2-x}Co_xTeO_6$ and Na_2Co_{2-}

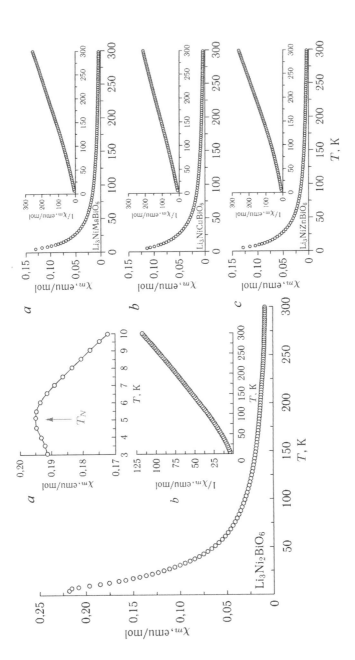

Fig. 6.15. Dependences of the magnetic susceptibility for Li$_3$Ni$_2$BiO$_6$ (the left panel) and for solid solutions of Li$_3$NiMgBiO$_6$ (*a*), Li$_3$NiCuBiO$_6$ (*b*), and Li$_3$NiZnBiO$_6$(*c*). The low-temperature AFM ordering in Li$_3$Ni$_2$BiO$_6$ is confirmed on the upper inset (on the left). The other insets show the temperature dependence of the reciprocal susceptibility [299]

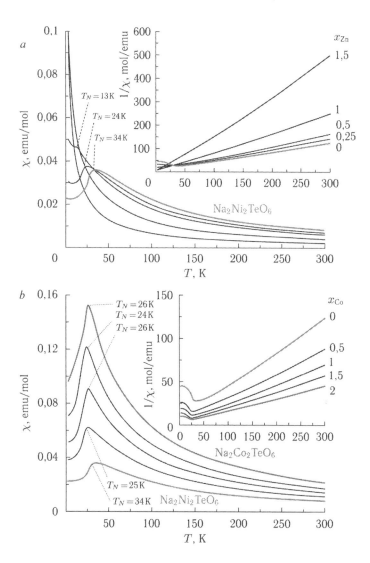

Fig. 6.16 *a*, *b*. Evolution of the magnetic susceptibility in the variation of the Ni concentration in $Na_2Ni_{2-x}Zn_xTeO_6$ (*a*) and $Na_2Ni_{2-x}Co_xTeO_6$ (*b*)

$_xZn_xTeO_6$ tellurates with partial or complete replacement of magnetic ions by non-magnetic ones have been carried out. Figure 6.16 shows the temperature dependence of the magnetic susceptibility of polycrystalline samples of solid solutions of $Na_2Ni_{2-x}Zn_xTeO_6$ (*a*) $Na_2Co_{2-x}Zn_xTeO_6$ (*b*) and $Na_2Ni_{2-x}Co_xTeO_6$ (*c*). For all samples in the high-temperature region, the inverse magnetic susceptibility follows a linear law, indicating the fulfillment of the Curie–Weiss law in the paramagnetic region. As nickel is replaced by non-magnetic zinc

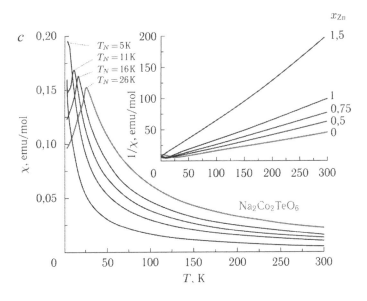

Fig. 6.16 *c*. Evolution of the magnetic susceptibility in the variation of Co concentration in $Na_2Co_{2-x}Zn_xTeO_6$ [297].

in the solid solution of $Na_2Ni_{2-x}Zn_xTeO_6$, the absolute values of the magnetic susceptibility and effective moment decrease.

The antiferromagnetic transition in $Na_2Co_{2-x}Zn_xTeO_6$ solid solution exists in a wider range of compositions than in the case of $Na_2Ni_{2-x}Co_xTeO_6$: up to $x = 1$ and $x = 0.5$, respectively. In the solid solution of $Na_2Ni_{2-x}Co_xTeO_6$, antiferromagnetic ordering was observed in a wide range of compositions with practically constant Néel temperature. The $\chi(T)$ curves for different compositions, shown in Fig. 6.16, demonstrate a decrease in the magnetic response when nickel and cobalt are replaced with zinc. The antiferromagnetic transition in doping with zinc moves to the region of lower temperatures. The paramagnetic character in the high-temperature region is confirmed for all samples by the linear inverse magnetic susceptibility approximated by the Curie–Weiss law in the temperature range 100–300 K. The effective magnetic moment increases with the replacement of nickel by cobalt, and the obtained value coincides with the theoretical value when only the spin contribution is taken into account. For all compositions of solid solutions, the Weiss temperature is negative, which indicates the predominance of antiferromagnetic interactions. However, the behaviour of the Weiss temperature is different when the composition of the solid solution for the nickel–zinc and cobalt–zinc tellurates varies: for $Na_2Ni_{2-x}Zn_xTeO_6$, it gradually increases to zero with increasing zinc concentration, and for $Na_2Co_{2-x}Zn_xTeO_6$, the Weiss temperature remains practically unchanged.

For most magnetically ordered antimonates and tellurates, anomalies in the magnetization curves induced by the magnetic field and, in particular, spin-reorientation transitions have been observed. For example, Fig. 6.17 *a–c* show the field dependences

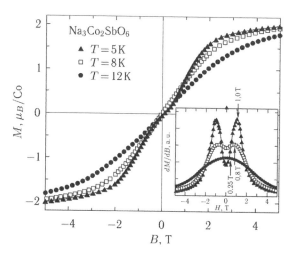

Fig. 6.17 *a*. Field dependences of the magnetization for $Na_3Co_2SbO_6$ [295]. The inset shows the derivatives *dM/dB*.

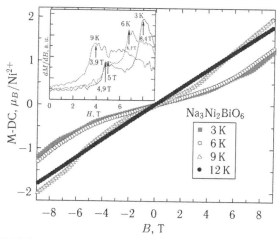

Fig. 6.17 *b*. Field dependences of the magnetization for $Na_3Ni_2BiO_6$ [301]. The inset shows the derivatives *dM/dB*.

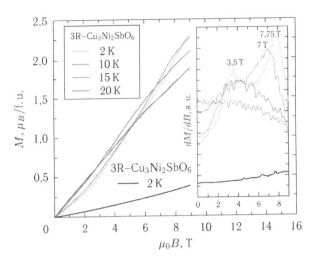

Fig. 6.17 *c*. Field dependences of the magnetization for $Cu_3Co_2SbO_6$ [302]. The inset shows the derivatives *dM/dB*.

of the magnetization for $Na_3Co_2SbO_6$ [295], $Na_3Ni_2BiO_6$ [301], and $Cu_3Co_2SbO_6$ [302]. It is interesting to note that in a number of cases, for example, in $Na_3Ni_2BiO_6$ [301] and $Cu_3Co_2SbO_6$ [302] two and more consecutive transitions, which indicates the presence of several phase boundaries on the magnetic phase diagram, which, apparently, correspond to the realization of different spin-configuration states on a honeycomb-type lattice.

Studies by neutron diffraction showed that the spin model in these compounds can be quite non-trivial. For example, it has been established that a zigzag antiferromagnetic order is realized in the $Na_3Ni_2BiO_6$ [301] system and the related delafossites $Cu_3Ni_2SbO_6$ and $Cu_3Co_2SbO_6$ [300], similar to the sodium irridate Na_2IrO_3 described above [271]. According to the proposed spin-configuration model (Fig. 6.18) [300], the main spin exchange in the layers proceeds along zigzag one-dimensional ferromagnetic chains of the edge-shared the octahedrons MO_6, which interact antiferromagnetically. Interestingly, for both nickel systems, $Na_3Ni_2BiO_6$ [301] and $Cu_3Ni_2SbO_6$ [300], the spins are aligned perpendicular to the plane of the honeycomb cells in accordance with uniaxial anisotropy, whereas in the cobalt sample the spins prefer orientation in the plane of the honeycomb cells.

For a number of layered tellurates and antimonates with a Jahn-Teller copper ion, $Cu^{2+}(S = 1/2)$, in particular, for $Na_3Cu_2SbO_6$ [303, 307, 311], $Na_2Cu_2TeO_6$ [307–311] and Cu_5SbO_6 [312], a spin-gap behaviour was observed, and the ground quantum state was

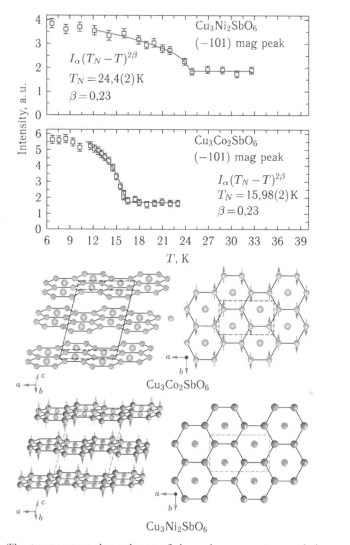

Fig. 6.18. The temperature dependence of the order parameters and the magnetic structure of Cu₃Co₂SbO₆ and Cu₃Ni₂SbO₆ [300].

determined as a spin singlet [311]. The temperature dependence of the magnetic susceptibility for these compounds demonstrates a wide maximum (Fig. 6.19), typical of a low-dimensional magnetic subsystem, which is replaced by an increase in $\chi(T)$ at low temperatures, which is caused by the presence of a small amount of paramagnetic impurity or a signal from chain interruptions. The Jahn–Teller distortion of the structure leads to a change in the nature of the magnetic interaction, causing a quasi-one dimensional character of the exchange, instead of the quasi-two dimensional interaction expected

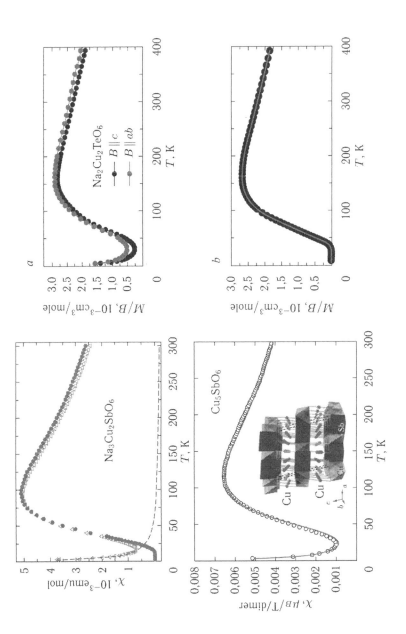

Fig. 6.19. The temperature dependences of the magnetic susceptibility of $Na_3Cu_2SbO_6$ [306]. $Na_2Cu_2TeO_6$ (a – for two directions of the magnetic field. b – is the shape of the curve after subtraction of the paramagnetic storage of impurities/defects) [310] and Cu_5SbO_6 [312].

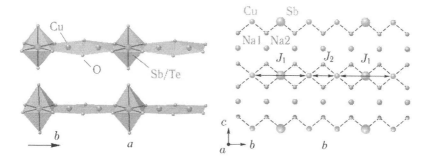

Fig. 6.20. Schematic representation of the crystal structure of the compound $Na_3Cu_2SbO_6$ (or $Na_2Cu_2TeO_6$), illustrating the Jahn-Teller distortion (*a*) [311] and the scheme of exchange interactions in $Na_3Cu_2SbO_6$ (*b*) [306]. The long arrow illustrates the super-superexchange interaction of the J_1 spins along the Cu–O–Sb–O–Cu path, and the short arrow shows the superexchange interaction of J_2 through Cu–O–Cu.

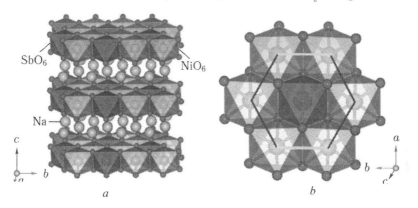

Fig. 6.21. Left panel: a polyhedral view of the layered crystalline structure of $Na_3Ni_2SbO_6$. Right panel: a fragment of the structure in the *ab* plane, showing the organization of non-equivalent Ni–Ni bonds in the magnetoactive layer.

for the layered structure (Fig. 6.20). Monte Carlo calculations show that for cuprate systems $Na_3Cu_2SbO_6$ and $Na_2Cu_2TeO_6$, the dominant exchange is antiferromagnetic and proceeds along a linear chain of Cu–O–Te (Sb)–O–Cu [311]. The model of the alternating AFM-FM single-dimensional chain satisfactorily describes the magnetic properties of these compounds.

6.5.1. Nickel antimonates

The structure of nickel antimonates $Na_3Ni_2SbO_6$, $Li_3Ni_2SbO_6$ is a superstructure from a lattice of rock salt with cationic ordering over layers [315, 316]. The lattice for both crystals is monoclinic, the

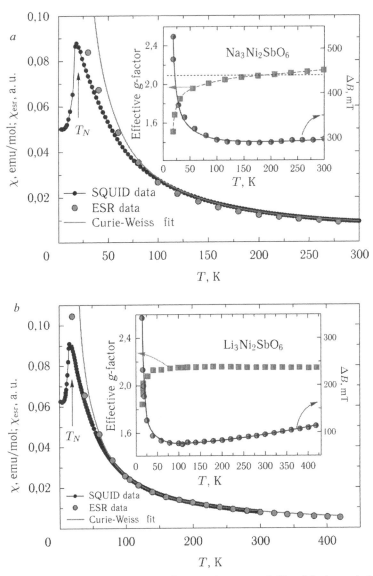

Fig. 6.22. Temperature dependences of magnetic susceptibility (black symbols) and integral ESR intensity (green symbols) for $Na_3Ni_2SbO_6$ (*a*) and $Li_3Ni_2SbO_6$ (*b*). The solid lines are the result of the approximation of the experimental data according to the Curie–Weiss law. On the insets: the temperature dependences of the effective g-factor (squares) and the width of the ESR line (circles). The solid lines on the inserts are the result of approximating the width of the ESR line in the framework of the theory of critical broadening.

space group is $C2/m$ (Fig. 6.21). In this structure, the magnetically active mixed layers of nickel and antimony cations in the octahedral oxygen environment alternate with non-magnetic layers of sodium (or lithium). In the magnetoactive layer, nickel cations are located in the edge-shared octahedra NiO_6, which are ordered into the honeycomb cell surrounding the antimony cation (right panel in Fig. 6.21).

An investigation of the magnetic susceptibility of nickel antimonates in a field of $B = 0.1$ T revealed that for both compounds the $\chi(T)$ dependence exhibits an acute maximum at ~ 15 K and ~ 17 K for lithium and sodium samples, respectively, and with a further decrease in temperature, the susceptibility decreases by $1/3$ (Fig. 6.22) [315, 316]. This behaviour indicates the establishment of a long-range antiferromagnetic order in matter at low temperatures and is typical of easy axis polycrystalline antiferromagnets. With an increase in the magnetic field strength, the maximum on the temperature dependences of the magnetic susceptibility broadens and shifts toward low temperatures.

In the temperature range above 100 K, the temperature dependence of the magnetic susceptibility is satisfactorily described by the Curie–Weiss law. The Weiss temperature then turns out to be positive and amounts to $\Theta \sim 8$ K and $\Theta \sim 12$ K for $Li_3Ni_2SbO_6$ and $Na_3Ni_2SbO_6$, respectively, which indicates the dominant ferromagnetic interaction between nickel ions Ni^{2+} in the magnetoactive layer. The effective magnetic moment, estimated from the Curie constant, is 4.3 and $4.4\mu_B$/f.u. for $Li_3Ni_2SbO_6$ and $Na_3Ni_2SbO_6$, which agrees well with the estimate $\mu_{theor} = 4.3\mu_B$/f.u., obtained for the g-factor $g = 2.15$.

An investigation of the ESR absorption spectra reveals the presence of a single broadened Lorentz-type line corresponding to a signal from Ni^{2+} ions in an octahedral oxygen environment. Evolution of the ESR absorption spectra in $Li_3Ni_2SbO_6$ with a temperature variation is shown in Fig. 6.23. It is seen that with a decrease in temperature, the line first narrows, and at $T < 100$ K it becomes noticeably broadened. In the region of the lowest temperatures, the line is considerably broadened, its amplitude decreases and, in the end, the ESR signal degrades, indicating the establishment of a long-range magnetic order.

The main parameters of the ESR spectra are determined by approximation using the Dyson-type function (4.1). On the whole, the character of the temperature dependences of the main ESR parameters is practically identical for both compounds (the insert in Fig. 6.22). From Fig. 6.22 it can be seen that in the paramagnetic phase χ_{ESR} varies in accordance with the Curie–Weiss law and agrees with the data on the static magnetic susceptibility for both samples. The average value of the effective g-factor $g = 2.15$ remains practically unchanged in the paramagnetic region up to ~ 140 K for the sample with Na and ~ 70 K for the sample with Li. Then, as the Néel temperature is approached, an appreciable shift of the resonant field to the region of large fields is observed. This behaviour corresponds to the presence of short-range correlations at temperatures much higher than the magnetic ordering temperature. The range of such correlations is much wider for the sodium sample, which is also characterized by a larger width of the absorption line. As a consequence, the value of the asymmetry parameter is $\alpha \sim 0.4$ for a sample with Na, while it is negligible, $\alpha \sim 0$, for a sample with Li. Note that $\alpha = 0$ corresponds to a symmetric Lorentz line, whereas $\alpha = 1$ corresponds to a completely asymmetric line, when absorption and dispersion are present in equal proportions.

The width of the absorption line decreases weakly with decreasing temperature linearly, passes through a minimum at ~ 140 K for a sample with Na and ~ 120 K for a sample with Li, and with a further decrease in temperature, it begins to increase rapidly. In the framework of the Mori–Kawasaki–Huber theory of critical broadening of the absorption line, when approaching T_N and taking into account the linear increase in the linewidth in the high-temperature region, we can describe the variation in the width of the ESR line ΔB in the entire temperature range studied with the help of expression (2.15) with the additional term $C \cdot T$:

$$\Delta B(T) = \Delta B^* + A \cdot \left[\frac{T_N^{ESR}}{T - T_N^{ESR}} \right]^{\beta} + C \cdot T. \qquad (6.3)$$

The best agreement with the experimental data was obtained for the values of the model parameters presented in Table 6.1.

Two important facts should be noted, which follow from an analysis of the behaviour of the width of the absorption line. First, a linear decrease and passage through the minimum was observed with decreasing temperature over a wide interval in the paramagnetic region. This kind of behaviour was observed experimentally for many 2D antiferromagnets [317–324], and the linear dependence of $\Delta B(T)$ was usually attributed either to phonon modulation of exchange interactions or to the influence of the crystal field (the latter for systems with $S > 1/2$). For transition metals, which in most the

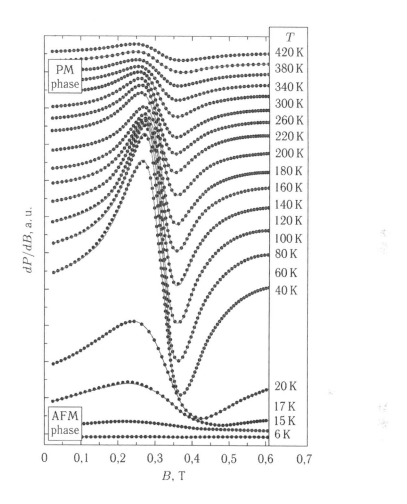

Fig. 6.23. Evolution of the ESR spectrum for $Li_3Ni_2SbO_6$ at a temperature variation.

Table 6.1. The parameters obtained from an analysis of the temperature dependences of the width of the ESR line in accordance with Eq. (6.3) for samples $A_3Ni_2SbO_6$ (A = Li, Na)

Sample synthesis method	T_N^{ESR}, K	ΔB^*, mT	A, mT	β	C, mT/K
$Li_3Ni_2SbO_6$ ion exchange	12 ± 1	70 ± 5	75 ± 5	1.2 ± 0.1	0.13
$Li_3Ni_2SbO_6$ solid-phase reaction	13 ± 1	25 ± 5	95 ± 5	0.8 ± 0.1	0.12
$Na_3Ni_2SbO_6$ solid-phase reaction	15 ± 1	230 ± 5	130 ± 5	0.5 ± 0.1	0.18

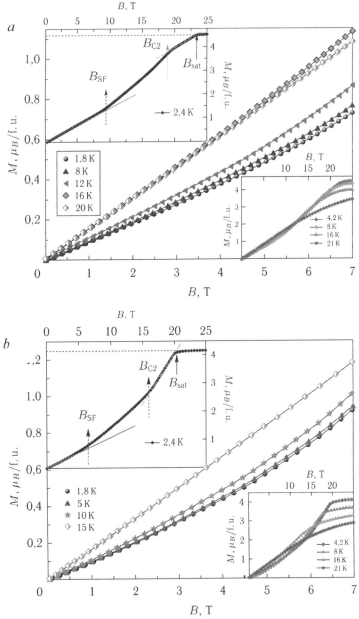

Fig. 6.24. Isotherms of magnetization in static and pulsed (inserts) magnetic fields for $Na_3Ni_2SbO_6$ (*a*) and $Li_3Ni_2SbO_6$ (*b*) at various temperatures.

orbital angular momentum is frozen, the effect of the crystal field is small and leads to $d(\Delta B)/dT \leq 0.1$ mT/K. In this case, however, the velocity $d(\Delta B)/dT$ is higher (see Table 6.1), which indicates an appreciable role of the orbital contribution for Ni^{2+} ions.

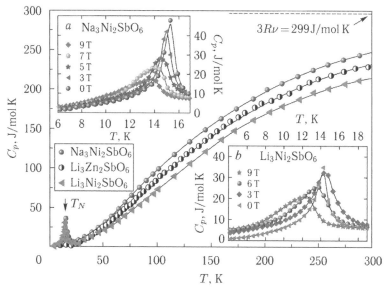

Fig. 6.25. Temperature dependences of specific heat capacity of $Li_3Ni_2SbO_6$, $Na_3Ni_2SbO_6$ and their non-magnetic analogue $Li_3Zn_2SbO_6$. Inserts: low-temperature parts of $C_p(T)$, showing the shift of T_N in magnetic fields.

The second point is that, within the Kawasaki theory, the critical exponent can be expressed as $\beta = [1/2 (7 + \eta) \nu - 2 (1 - \alpha)]$, where ν describes the divergence of the correlation length, η is the critical exponent for the divergence of static correlations and α refers to the divergence of the specific heat, respectively. For the 3D Heisenberg antiferromagnet $\eta = \alpha = 0$ and $\nu = 2/3$ [103, 105], then the critical exponent $\beta = 1/3$, which is noticeably lower than the values found (see Table 6.1). Thus, the analysis of the spin dynamics of the compounds $A_3Ni_2SbO_6$ (A = Li, Na) testifies to the 2D character of the exchange interactions.

The field dependences of the magnetization in static and pulsed fields for $A_3Ni_2SbO_6$ (A = Li, Na) are shown in Fig. 6.24. Saturation of the magnetic moment is observed in the fields $B_{sat} \sim 23$ and 20 T for $Na_3Ni_2SbO_6$ and $Li_3Ni_2SbO_6$, respectively, and M_{sat} is in good agreement with the theoretically expected saturation moment of 4.3 μ_B/f. u. (the upper inserts in Fig. 6.24). Field dependences show a change in the slope with increasing applied field, characteristic for the spin-reorientation transition with critical B_{SF} fields of 9.8 and 5.5 T for $Na_3Ni_2SbO_6$ and $Li_3Ni_2SbO_6$, respectively.

Data on the specific heat for samples $A_3Ni_2SbO_6$ (A = Li, Na) and diamagnetic analog $Li_3Zn_2SbO_6$ are shown in Fig. 6.25. The

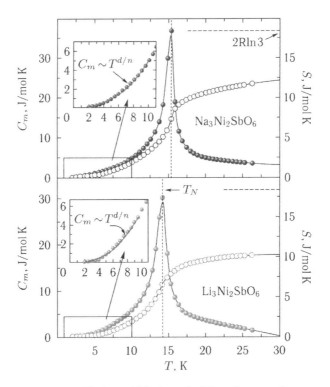

Fig. 6.26. Magnetic specific heat (filled symbols) and magnetic entropy (empty symbols) for $A_3Ni_2SbO_6$ (A = Li, Na). Inserts: low-temperature parts and their approximation.

$C_p(T)$ dependences are in agreement with the data on the magnetic susceptibility in weak magnetic fields and show distinct λ-type anomalies, characteristic for establishing the long-range magnetic order. In the magnetic field, the λ-type anomalies broadens and shifts toward lower temperatures, indicating a shift in the phase boundary (the insert in Fig. 6.25). Analysis of the magnetic contribution of C_m to the heat capacity of samples $A_3Ni_2SbO_6$ (A = Li, Na) below the Néel temperature within the framework of the spin-wave theory $C_m \sim T^{d/n}$ gives the values of $d = 3$ and $n = 1$ for both (Li and Na) samples in Fig. 6.26), which confirms the 3D AFM magnon's picture at low temperatures.

It can be seen that the magnetic entropy is saturated at a temperature above 25 K, reaching ~ 10 J/mol K (Fig. 6.26), which is noticeably lower than the value obtained from the average field theory estimate of 18.3 J/mol K. Only ~ 40% entropy is released below the temperature T_N, which indicates the correlation of the short-range order at $T \gg T_N$.

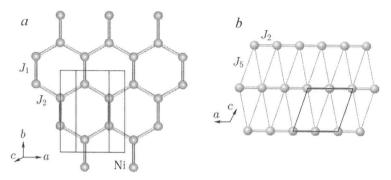

Fig. 6.27. The main ways of exchange interactions in $Na_3Ni_2SbO_6$: *a*) exchanges between nearest neighbors in a hexagon of a magnetoactive layer: the exchange at the shortest distance J_1 is antiferromagnetic, the exchange between neighbors at a longer distance J_2 is ferromagnetic; *b*) the exchange of J_5 between magnetically active layers is a weak antiferromagnetic.

Table 6.2. The parameters of exchange interactions in $A_3Ni_2SbO_6$ (A = Li, Na)

	J_1, K	J_2, K	J_3, K	J_5, K
$Na_3Ni_2SbO_6$	15	−22	-----	1
$Li_3Ni_2SbO_6$	18	−25	-----	2

As was established above from the analysis of the temperature dependence of the magnetic susceptibility of $Na_3Ni_2SbO_6$ and $Li_3Ni_2SbO_6$, the Weiss temperature was positive, which indicates the dominance of the ferromagnetic exchange interaction, despite the establishment of a long-range antiferromagnetic order. In order to determine the character and sign of the exchange interactions, as well as the most probable spin-configuration model of the magnetic structure, calculations were performed from the first principles.

When analyzing the magnetic interactions in $Na_3Ni_2SbO_6$ and $Li_3Ni_2SbO_6$, it was taken into account that there is a slight difference in the lengths of the cation-cation bonds between the nearest neighbors in a honeycomb cell type. In $Na_3Ni_2SbO_6$, there are two distances d_{Ni-Ni} = 305 Å and four distances d_{Ni-Ni} = 306 Å as shown on the right panel of Fig. 6.21. For $Li_3Ni_2SbO_6$, these values are d_{Ni-Ni} = 298 Å and d_{Ni-Ni} = 299 Å, respectively.

The main exchange integrals J_1 and J_2 in the layer and between the layers J_5 (the notations correspond to those shown in Fig. 6.27) are presented in Table 6.2. It was established that the exchange within the magnetically active layer of honeycomb cells between the two nearest neighbors Ni_i and Ni_{i+1} J_1 is antiferromagnetic, while the exchange between other neighbors Ni_i and Ni_{i-1} J_2 is ferromagnetic (Fig. 6.27). In this case, the exchange between neighbors following the closest in the hexagon, J_3 is antiferromagnetic and very small. The exchange between the magnetic layers J_5 is also antiferromagnetic, but an order of magnitude smaller than for intra-layer exchanges. The results of calculations for various spin-configuration models have shown that in the ground state a zigzag magnetic order is established, the energy of which is lower than the energy of the ferromagnetic order.

Comparison of the calculation results for the two systems studied shows that the main exchange integrals J_1 and J_2 take significantly larger values for $Li_3Ni_2SbO_6$, which is due to shorter distances of Ni–Ni in the lithium compound. It should be noted that the decrease in the distance between the magnetoactive layers in the lithium sample, due to the smaller ionic radius of lithium ions compared to sodium, leads to a greater value of the interlayer exchange J_5.

Using an estimate of the energy of the magnetic crystallographic anisotropy obtained from theoretical calculations and the value $\Delta = 2.4$ meV/f. u. energy gap between the ferromagnetic and zigzag antiferromagnetic states, the spin-flop field can be estimated within the framework of the approach [325]: $B_{SF} = 2\sqrt{K}\Delta / M$, where K is the energy of the magneto-crystallographic anisotropy, and M is the magnetic moment of Ni. The obtained estimate of the spin-flop field $B_{SF} = 7.2$ T for $Na_3Ni_2SbO_6$ is lower than the experimental value (9.8 T), but comparable with it in order of magnitude. An estimate of the saturation field yields a value of about 20 T in agreement with the experimental value of $B_{sat} \sim 23$ T.

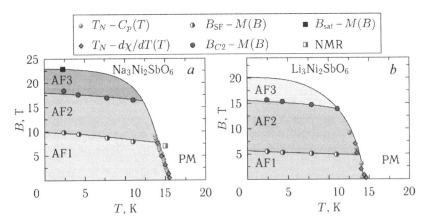

Fig. 6.28. Magnetic phase diagrams for antimonates $A_3Ni_2SbO_6$. Left panel A = Na; right panel A = Li.

Based on the results of thermodynamic and resonance studies, magnetic phase diagrams were constructed for layered antimonates of nickel and alkali metals $A_3Ni_2SbO_6$ (A = Na, Li), shown in Fig. 6.28. At temperatures above the Néel temperature (16 K for the sodium and 14 K for the lithium sample, respectively), a paramagnetic phase is realized in the zero magnetic field. As the value of the magnetic field increases, the phase boundary shifts toward low temperatures. The antiferromagnetic phase, however, is complicated by the presence at low temperatures of two additional phases induced by the magnetic field, which, apparently, arise due to the reorientation of spins in a magnetic field. Thus, the investigated samples can be classified as quasi-two-dimensional, easy-axis antiferromagnets with a ground state of zigzag antiferromagnetic (AF1) type. The spin-flop-induced magnetic field phase (AF2) is realized in the fields 5–15 T for $Li_3Ni_2SbO_6$ and 10–18 T for $Na_3Ni_2SbO_6$, and with further increase of the magnetic field strength it is replaced by a phase of another type (AF3), probably corresponding to another spin configuration. The obtained phase diagrams agree satisfactorily with theoretical calculations from the first principles, which show that in both investigated nickel antimonates, a quasi-2D antiferromagnetic zigzag order is established (ferromagnetic zigzag chains connected antiferromagnetically in magnetically active layers) with a preferred orientation of the spins perpendicular to the plane of the honeycombs.

6.5.2. Cobalt antimonates

Cationic ordering along the layers in the crystal structure of $Na_3Co_2SbO_6$ and $Ag_3Co_2SbO_6$ is completely analogous to the isostructural antimonates $A_3Ni_2SbO_6$ (A = Li, Na) with a magnetic subsystem of cobalt ions on a lattice of the honeycomb-type [326]. The lattice in the magnetoactive layer is formed by edge-shared CoO_6 octahedra, however, the Co–Co bonds are not equivalent in length, as shown by the lines in the right panel of Fig. 6.29. The arrangement of nonmagnetic layers, however, is different, since for an argentum sample the monovalent Ag^+ cations are in linear oxygen coordination, and accordingly the $Ag_3Co_2SbO_6$ antimonate is characterized by a delafossite type structure (Fig. 6.29).

It is established that both cobalt systems studied are ordered antiferromagnetically (Fig. 6.30), and in the high-temperature region the $\chi(T)$ dependences are satisfactorily described by the Curie–Weiss law with a negative Weiss temperature of $\Theta \sim -10$ K for

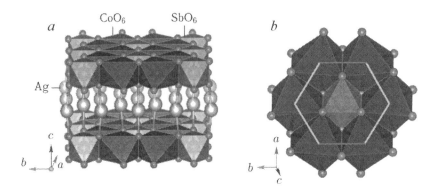

Fig. 6.29. The general appearance of the layered crystal structure in $Ag_3Co_2SbO_6$ (*a*) and the polyhedral type of honeycomb-type cell (*b*).

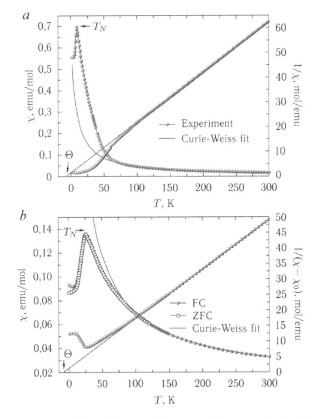

Fig. 6.30. Temperature dependences of the magnetic susceptibility for $Na_3Co_2SbO_6$ (*a*) and $Ag_3Co_2SbO_6$ (*b*) and the inverse magnetic susceptibility in the field $B = 0.1$ T. Solid lines – approximation in accordance with the Curie–Weiss law.

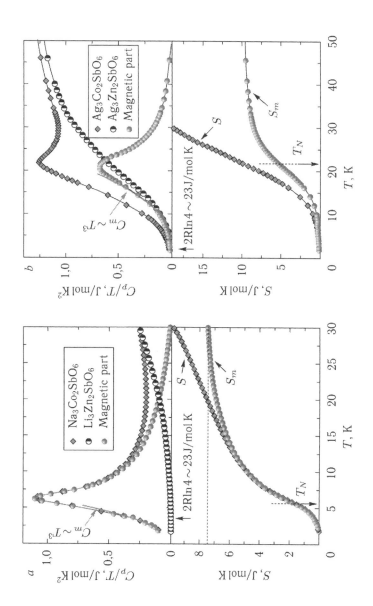

Fig. 6.31. Temperature dependences of the specific heat and entropy in $Na_3Co_2SbO_6$ (*a*) and $Ag_3Co_2SbO_6$ (*b*).

both compounds [326]. At $T < 100$ K, the $\chi(T)$ dependences deviate noticeably from the Curie–Weiss behaviour, indicating the presence of ferromagnetic exchanges in $Na_3Co_2SbO_6$ and the enhancement of antiferromagnetic exchanges in $Ag_3Co_2SbO_6$. The effective magnetic moment $\mu_{eff} = 6.7\mu_B$/f.u. in $Ag_3Co_2SbO_6$ is in good agreement with the theoretical value obtained using the g-factor $g = 2.4$. At the same time, $\mu_{eff} = 6.5\mu_B$/f.u. in $Na_3Co_2SbO_6$ is below the theoretical value at the experimentally established value $g = 3.3$. In magnetic fields, the Néel temperature in the argentum sample is slightly shifted toward lower values, whereas in the sodium sample the character of the $\chi(T)$ dependences changes drastically, demonstrating a rapid suppression of antiferromagnetism in the field of ~ 1.5 T.

The data on specific heat, shown in Fig. 6.31, agree with the data on the magnetic susceptibility and indicate the establishment of a long-range magnetic order at low temperatures. The $C_p(T)$ dependences in the zero field have an λ-type anomaly, which makes it possible to estimate the Néel temperature as $T_N \sim 6.7$ K and 21.2 K for $Na_3Co_2SbO_6$ and $Ag_3Co_2SbO_6$, respectively. To estimate the magnetic contribution to the heat capacity and entropy, we used diamagnetic analogues $Li_3Zn_2SbO_6$ and $Ag_3Zn_2SbO_6$.

It is also established that the values of the jumps in the heat capacity and the level of saturation of the magnetic entropy (Fig. 6.31) are much smaller than those estimated from the mean-field theory. This indicates a large contribution of short-range order correlations over a wide temperature range above the transition to a magnetically ordered state for both compounds. Analysis of the magnetic heat capacity $C_m(T)$ below the Néel temperature within the framework of the spin-wave theory in accordance with the power law gives $d = 3$ and $n = 1$ for Na and Ag samples with good accuracy, which confirms the 3D AFM magnon pattern at low temperatures.

The magnetization isotherms for both compounds exhibit an anomaly at low temperatures, indicative of a spin-flop transition, as shown in Fig. 6.32. With increasing temperature, the position of the anomalies shifts toward smaller ($NaCo_2SbO_6$) or larger ($AgCo_2SbO_6$) fields, then at $T > T_N$ the anomalies on $M(B)$ vanish. Based on the results of thermodynamic measurements, the magnetic phase diagrams in the magnetic field–temperature coordinates for $A_3Co_2SbO_6(A = $ Na, Ag) shown in Fig. 6.33.

Calculations from the first principles were carried out to determine the nature and sign of the exchange interactions, as well as the establishment of the spin-configuration model of the magnetic

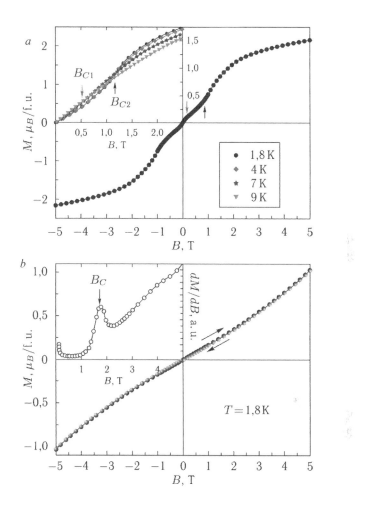

Fig. 6.32. Isotherms of magnetization in $Na_3Co_2SbO_6$ (*a*) and $Ag_3Co_2SbO_6$ (*b*). The arrows indicate the critical fields.

structure. The spin moment on the Co ion was $2.7\mu_B$, which is close to the ion value for Co^{2+}, and a small reduction is associated with the typical values for the transition elements effects of hybridization. The main ways of intra-layer exchange interactions are shown in the bottom panel of Fig. 6.34.

It is found that the magnetic structure of the quasi-two-dimensional AFM of the zigzag type in the quantum ground state is the most energetically favourable for the argentum sample of $Ag_3Co_2SbO_6$ in the main quantum state: ferromagnetic zigzag chains are antiferromagnetic in the magnetically active layers (the lower

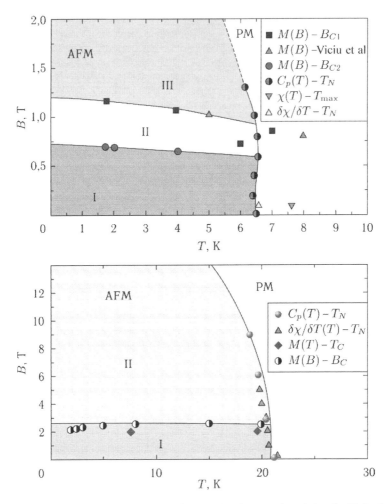

Fig. 6.33. Magnetic phase diagrams for $Na_3Co_2SbO_6$ (top) and $Ag_3Co_2SbO_6$ (bottom). I, II, III – AFM phases with different spin configurations.

panel of Fig. 6.34). The main exchange integrals were: $J_1^s = 28$ K (AFM), $J_1^l = -2$ K (FM), and $J_2 = 3$ K (AFM). An unexpected result for the cobalt antimonates was a large difference between ferromagnetic J_1^l and antiferromagnetic J_1^s exchanges between the nearest neighbors in a honeycomb-type cell, in contrast to the nickel samples, where these exchanges are comparable in magnitude (see the parameters given in Table 6.2). The difference in the interatomic distances for long and short Co–Co contacts is very small, similar to the situation of Ni–Ni contacts in $A_3Ni_2SbO_6$.

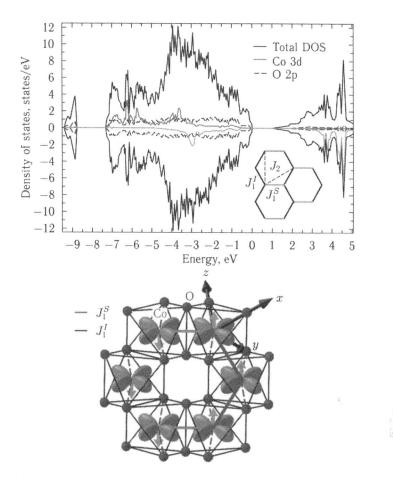

Fig. 6.34. Top panel: total and partial density of states functions in $Ag_3Co_2SbO_6$. On the inset: the main ways of intralayer exchange interactions. The bottom panel: the inhabited order (one half-filled t_{2g} Co orbital). AFM zigzag is shown by arrows.

In order to find the mechanism responsible for this difference, the features of the orbital structure were analyzed. As noted above, the Co ions are in an octahedral environment in $Ag_3Co_2SbO_6$. In addition, these octahedra turn out to be compressed, which leads to a certain splitting of the $3d$-shell Co. If one selects a local coordinate system, as shown in the lower panel of Fig. 6.34 (i.e., all axes are directed at the oxygen ions), then the t_{2g} group of orbitals is divided into the lower xy orbitals and higher in energy zx/yz orbitals. In the $3d^7$ configuration of Co^{2+}, the sub-band zx/yz turns out to be filled by 3/4 and, consequently, is magnetoactive. The peculiarity of the $Ag_3Co_2SbO_6$ crystal structure that the short Co–O bonds are directed differently in CoO_6 pairs of octahedra in short and long Co–Co contacts, leads to orbital ordering, which causes significantly different exchange parameters for these bonds. Theoretical calculations were made to construct a single half-filled t_{2g} orbital. It was found that as a result of orbital ordering, the individual half-filled t_{2g} orbitals shown in Fig. 6.34 by the red 'eights', involved in the formation of short Co–Co bonds, are directed to each other (these are the $xz + yz$ orbitals in the coordinate system with the z axis along the shortest Co–O bond). Such an orbital order leads to the fact that the AFM exchange occurs both in direct and in super-exchange mechanism between the nearest octahedra of CoO_6.

If we consider the contact of two Co ions participating in the formation of long contacts (the lines in Fig. 6.34), the orbitals that are half filled with $xz + yz$ do not overlap with one another, i.e., the direct AFM exchange is suppressed due to the orbital order. In addition, the superexchange with the help of the same p-orbitals of oxygen is impossible because of the difference in signs of p- and d-wave functions. A superexchange with the participation of two different p-orbitals is possible, but it is weak and ferromagnetic [327]. A superexchange with the participation of e_g orbitals also turns out to be ineffective in the geometry of edge-shared octahedra along the edge [54, 328]. Thus, the analysis showed that the orbital order in $Ag_3Co_2SbO_6$ blocks the AFM t_{2g}–t_{2g} exchange between a part of the nearest neighbors in a hexagon from Co ions.

The spin-configuration model of $Na_3Co_2SbO_6$ was determined experimentally by neutron diffraction. The magnetic ordering is characterized by the presence of two propagation vectors, one of which is disproportionate: $k_1 = (1/2, 1/2, 0)$ and $k_2 = (0, 1/2 + \delta, 1/2)$ with $\delta \approx 0.011$. The spin structure is formed by a superposition of the collinear AFM (k_1) component in the general direction with a long-period sinusoidal (k_2) component along the a axis. The resulting spin model for $Na_3Co_2SbO_6$ assumes ferromagnetic zigzag chains that are antiferromagnetically bound in the honeycomb layers. These chains, however, are not collinear, as is the case in the isostructural antimonates $Ag_3Co_2SbO_6$ and $A_3Ni_2SbO_6$ (A = Li, Na).

Quasi-two-dimensional magnets with triangular motifs in the structure

7.1. Frustration of exchange interactions in the lattice of kagome and in diamond chains

Classic examples of systems with geometric frustration are two-dimensional lattices of kagome and diamond chains. The lattice of the type of diamond chains is based on diamond-shaped structural units. In contrast to the triangular lattice, the peculiarity of the diamond chain is that it is formed not from equilateral but from isosceles triangles, at the vertices of which magnetic ions are located, that is, both the geometric and exchange frustration of magnetic interactions is assumed. The left panel of Fig. 7.1 shows a variant of the arrangement of a diamond chain with three inequivalent exchange integrals. To form a two-dimensional grid, we need to consider a more complex picture of the interaction, namely, take into account either additional exchanges along the chain itself, or the interaction between the neighbouring chains.

The Hamiltonian of the problem for an isolated diamond chain is written as:

$$\hat{H} = J_1 \sum_{i=1}^{N/3} (\hat{S}_{3i-1}\hat{S}_{3i} + \hat{S}_{3i}\hat{S}_{3i+1}) + J_2 \sum_{i=1}^{N/3} (\hat{S}_{3i+1}\hat{S}_{3i+2}) + \\ + J_3 \sum_{i=1}^{N/3} (\hat{S}_{3i-2}\hat{S}_{3i} + \hat{S}_{3i}\hat{S}_{3i+2}),$$

(7.1)

where $\hat{S}$ is the spin operator for the moment $S = 1/2$, N is the total number of spins, and the exchange integrals J_1, J_2, J_3 are assumed to be antiferromagnetic (> 0) [329, 330]. Such a model is the most common, it is called a 'distorted diamond chain', although more often symmetric cases are often considered, for example $J_1 = J_3$ [331, 332].

Depending on the ratio of the exchange integrals in the diamond chain, one of the three ground states (at $B = 0$) can be realized: ferrimagnetic with the total angular momentum $S_{tot} = N/6$, and two spin-singlet states with $S_{tot} = 0$, or a system of non-interacting dimers, or a spin liquid [330].

The kagome lattice, shown in the central panel of Fig. 7.1, is formed from hexagons, which, in contrast to the honeycomb lattice, are displaced relative to each other for half-period. Thus, equilateral triangles with geometric frustration of the exchange interaction are formed in the structure. In the ideal model case, only one exchange integral J is considered in the system, and each magnetic ion interacts with the 4 nearest neighbors. Theoretical calculations by the method of diagonalizing the density matrix of states [333] indicate a spin-liquid ground state of such a system with a small spin gap of the order of 1/20 of the exchange interaction parameter. As the temperature is lowered, the kagome lattice becomes a non-magnetic ground state without a phase transition. The resonant valence state includes eight-link loops (octagons) and dimers, shown by dashes of different thickness in the right panel of Fig. 7.1. The lowest excitation energies of spinors in such a system have a spin $S = 0$ [334].

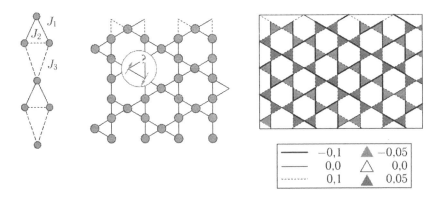

—— $-0,1$	▲ $-0,05$
—— $0,0$	△ $0,0$
------ $0,1$	▲ $0,05$

Fig. 7.1. Left panel: a diamond chain. In the centre: the arrangement of atoms in the kagome lattice. Right panel: the ground state of the kagome lattice $S = 1/2$ with spinon energies [333].

7.1.1. Herbertsmithite and vésigniéite

The best known metal oxide compound, where the model of the antiferromagnetic kagome lattice $S = 1/2$ is realized, is the mineral herbertsmithite $ZnCu_3(OH)_6Cl_2$. This compound has a rhombohedral lattice, the space group is $\bar{R}3m$. In this structure, the kagome layers are composed of $Cu_3(OH)_6$ triangles, as shown in Fig. 7.2. Zinc ions are surrounded by six OH^- hydroxyl groups between the magnetoactive layers. The Jahn–Teller Cu^{2+} cations are in a square oxygen environment. The X-ray diffraction method has shown that the Jahn–Teller position in the layer is occupied mainly by copper cations, whereas the interlayer position has a mixed Cu/Zn occupation [335].

The temperature dependence of the magnetic susceptibility of $Cu_4(OH)_6Cl_2$ on the temperature shows a ferrimagnetic ordering at $T_C = 4.5$ K, as shown in the inset at the bottom of Fig. 7.2. As the Zn content in in $Zn_xCu_{4-x}(OH)_6Cl$ increases, the ferrimagnetic transition is suppressed, while the Weiss temperature obtained from the treatment of the high-temperature $\chi(T)$ dependence is negative and increases in absolute value. In $ZnCu_3(OH)_6Cl_2$, no formation of long-range magnetic order is observed, and $\Theta = -314$ K [335].

The temperature dependence of the magnetic susceptibility of $ZnCu_3(OH)_6Cl_2$, contrary to expectations for the antiferromagnetic kagome lattice $S = 1/2$, increased somewhat with decreasing temperature, as shown in the bottom panel of Fig. 7.2. However, studies of the ^{35}Cl NMR spectra made it possible to separate the volume magnetic susceptibility proportional to $^{35}K_{1/2}$ and the intrinsic susceptibility of the matrix, which is proportional to the Knight shift of ^{35}K, as shown in Fig. 7.3. It is seen that below 50 K the susceptibility of the matrix decreases. The increase in volume susceptibility was attributed to the presence of impurity defect centres and to the Dzyaloshinsky–Moriya interaction [336].

Studies of the specific heat of $ZnCu_3(OH)_6Cl_2$ also produced an unexpected result. In [337], a broad maximum expected at low temperatures for a spin-gap system at low temperatures was found on the $C_p(T)$ curve, which was completely suppressed by a magnetic field of 14 T, as shown in Fig. 7.4. However, studies of the specific heat at ultralow temperatures have shown that the specific heat is described by the power function $C_p = \gamma T^\alpha$, and the exponent α varies: $\alpha = 1$ when processing results in the range up to 400 mK and $\alpha = 2/3$ in a wider range up to 600 mK, as shown by solid lines on the bottom panel of Fig. 7.4.

Inelastic neutron scattering spectra in $ZnCu_3(OH)_6Cl_2$ did not reveal indications of a spin gap up to $J/170$. The ground state model proposed by the authors is a two-dimensional spin liquid with nascent spinons (paired spin excitations) [337].

Another candidate for the realization of the quantum spin liquid state was the mineral vésiginité $BaCu_3V_2O_8(OH)_2$ structurally related to herbertsmithite, which, however, shows a very different type of magnetic behaviour [321, 338–343]. This compound crystallizes into a monoclinic $C2/m$ layer structure, with Cu^{2+} ions in a distorted octahedral coordination forming a kagome lattice, as shown in Fig. 7.5 *a* [343]. Unlike from herbertsmithite, vésiginité shows the

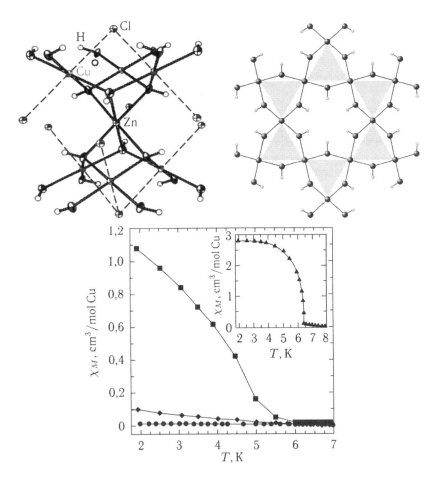

Fig. 7.2. Top panel: crystal structure of herbertsmithite $ZnCu_3(OH)_6Cl_2$. Bottom panel shows the temperature dependences of the magnetic susceptibility $Zn_xCu_{4-x}(OH)_6Cl_2$, $x = 0.5$, 0.7, 1. On the inset: the magnetic susceptibility of the initial compound $x = 0$ [335].

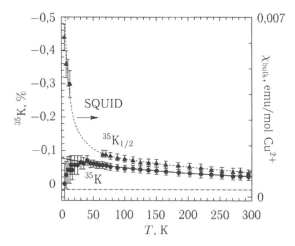

Fig. 7.3. The Knight shift for the smallest component ^{35}K of the distributed local susceptibility (triangles) and half the intensity of $^{35}K_{1/2}$ (circles) of the main peak in the ^{35}Cl NMR spectrum. The points are $-\chi(T)$, the dashed line is a contribution independent of temperature [336].

establishment of a long-range magnetic order of the Néel type ($q =$ 0) with $T_N = 9$ K. In this case, the dependence $\chi(T)$ shown in Fig. 7.5 *b*, has a sharp maximum. It is established that the copper ions form a non-collinear 120° spin configuration, and magnetism is determined by the dominant antiferromagnetic exchange between the nearest neighbors $J = 53$ K. Investigations by the NMR method [342, 343] have shown the essential role of the Dzyaloshinsky–Moriya anisotropy for the clarification of which the EPSR method was used [321]. Analysis of the temperature dependences of the main EPSR parameters, as shown by the solid lines in Fig. 7.6, made it possible to estimate the contribution of the Dzyaloshinsky–Moriya interaction. It is established that a large intrasplane anisotropy effectively suppresses quantum spin fluctuations in $BaCu_3V_2O_8(OH)_2$, which leads to stabilization of the long-range magnetic order in this compound, unlike herbertsmithite [321].

7.1.2. The combination of kagome and triangular layers in quasi-2D cobaltites

The combination of two-dimensional kagome and triangular layers containing magnetic ions was discovered and studied in detail on a number of compounds with the general formula $ABaCo_4O_7$, where A is a rare earth or alkaline earth metal or a combination thereof.

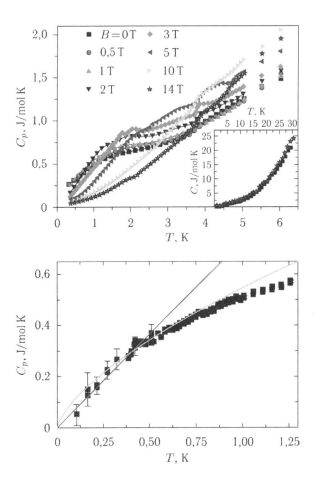

Fig. 7.4. Top panel: temperature dependences of specific heat of $ZnCu_3(OH)_6Cl_2$ in magnetic fields. Bottom panel: specific heat in a zero field at ultralow temperatures [337].

The structure of the initial compound $YBaCo_4O_7$ (by analogy with superconductors, the name '114' is often used) is hexagonal, the space group is $P6_3mc$ [344]. In the crystal structure of $YBaCo_4O_7$, the cobalt atoms are in the CoO_4 tetrahedra and are located in two crystallographic positions (Co1 and Co2 in Fig. 7.7). Tetrahedra connect at the corners and form a grid in the crystal, similar to the würtzite structure.

Two types of layers containing CoO_4 tetrahedra alternate along the c axis. In the layers of the first type, there are ions in the Co2 position (light tetrahedrons in Fig. 7.7), which form a kagome grid in the ab plane of the kagome. In the layers of the second type,

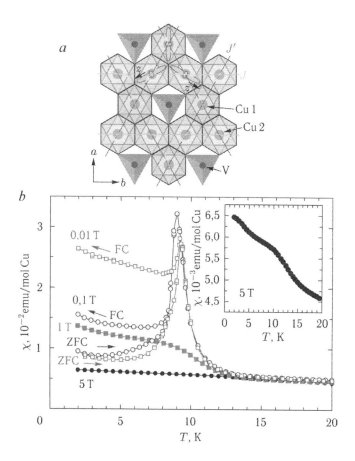

Fig. 7.5. The fragment of the crystal structure in the magnetoactive layer (*a*) and the temperature dependence of the magnetic susceptibility for the variation of the magnetic field (*b*) for BaCu$_3$V$_2$O$_8$(OH)$_2$ [343].

CoO$_4$ tetrahedra are in the Co1 position (dark tetrahedra in Fig. 7.7). The last layers are highly rarefied: the tetrahedra in them are not connected to one another, but the neighbouring layers containing Co2 bind [345].

Structural analysis showed that the tetrahedrons themselves are distorted and they contain a whole set of angles and metal–oxygen–metal distances. A characteristic feature of the structure of magnetic layers is the triangular motif in the structure. Such a device assumes a strong frustration of the magnetic interaction between the Co ions.

In the YBaCo$_4$O$_7$ compound, the Co^{2+}:Co^{3+} ratio is 3:1, however, charge ordering is not observed. It was initially thought that this compound belongs to the class of disordered magnets, since on the magnetic susceptibility there was observed a single anomaly

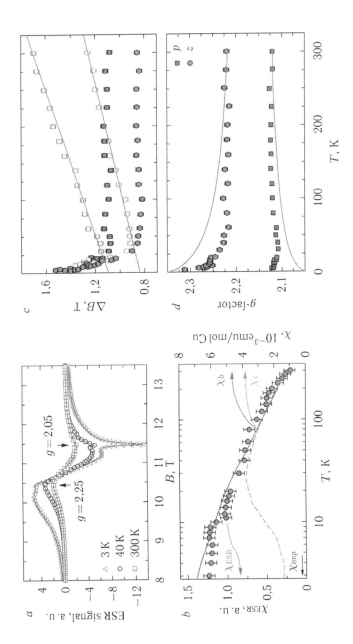

Fig. 7.6. The ESR signal (*a*) and the temperature dependences of the main ESR parameters: the integral intensity χ_{ESR} (*b*), the absorption line width $\Delta B(in)$ (*c*), the effective *g*-factor (*d*) for $BaCu_3V_2O_8(OH)_2$ [321].

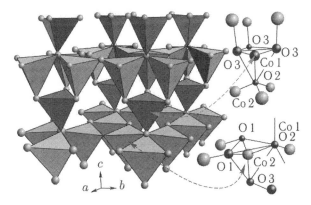

Fig. 7.7. Tetrahedra containing Co1 (dark) and Co2 (light) in the würtzite grid in YBaCo$_4$O$_7$.

corresponding to the 'freezing' of the magnetic subsystem at $T_f = 66$ K. In this case, the Weiss paramagnetic temperature is $\Theta \sim -900$ K, which corresponds to the frustration parameter $\Theta/T_f \sim 14$ [344]. Experiments on neutron scattering on single crystals have shown the formation of a regime of short-range correlations below 105 K (the appearance of diffuse neutron scattering), the results being treated as structural phase transitions at $T_{C1} \sim 70$ K and $T_{C2} \sim 105$ K for the layer with the kagome motif and the layer with triangles, respectively [346]. Further studies have shown that the magnetic properties of grown single crystals are significantly affected by stoichiometry in oxygen, and special efforts have been made to grow stoichiometric single crystals [347]. On the last samples, the symmetry group was refined and it was established that at $T_S = 313$ K a structural phase transition occurs from the trigonal high-temperature phase with the spatial symmetry group $P31c$ (2 positions for Co) to low-temperature orthorhombic $Pbn2_1$ with the superlattice $(a, \sqrt{3a}, c)$ (4 positions for Co).

At $T > T_S$, the behaviour of the magnetic susceptibility shown in Fig. 7.8 [347], is described by the Curie–Weiss law with a constant $C = 10.08$ emu K/mol and a Weiss temperature of $\Theta \sim -500$ K, which corresponds to a strong antiferromagnetic interaction. The evaluation of the effective magnetic moment in this temperature range is in good agreement with the ratio $3\text{Co}^{2+}(d^7):\text{Co}^{3+}(d^6)$. In the structural phase transition, the position of the oxygen ions changes, which is accompanied by a distortion of the tetrahedra in the kagome layer. However, the change in the character of the $\chi(T)$ dependence is due not to the distortion of the oxygen environment, but to the

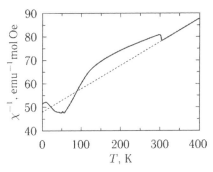

Fig. 7.8. The temperature dependence of the inverse magnetic susceptibility of $YBaCo_4O_7$ [347].

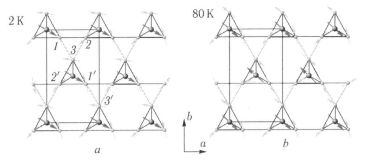

Fig. 7.9. Magnetic structure in $YBaCo_4O_7$ in the projection onto the *ab* plane. Dark arrows – Co1, light – Co2 [347].

formation of the short-range order regime, since with decreasing symmetry the frustration of the exchange interaction is eliminated. Magnetic correlations of the short-range order are seen in the neutron scattering pattern, in particular, diffuse scattering is observed below 250 K. At $T < 115$ K, a diffraction peak appears corresponding to the formation of a long-range magnetic order with a vector $k = 0$ in the orthorhombic symmetry of the unit cell (or $k = (1/2, 0, 0)$ in the hexagonal representation of the structure), as shown in Fig. 7.9 *b*. The magnetic susceptibility exhibits a maximum at $T \sim 80$ K. With a further decrease in temperature, a partial reorientation of the magnetic moments is observed. At $T = 2$ K, the magnetic structure shown in Fig. 7.9 *a*. It is seen that the magnetic moments lie mainly in the plane *ab*, and for the moments with the numbers 1–3 there are antiparallel pairs with the numbers 1'–3'.

In the layers containing triangles, the magnetic moment is 3.29 μ_B, which is close to the expected value for the spin moment of Co ions in the average oxidation state +2.25. In the layers with the kagome structure, the moment is much smaller ($2.19\mu_B$).

As a result of the structural phase transition, frustrations of exchange below 313 K are eliminated. All exchanges in the system are antiferromagnetic. In the kagome layer there are two types of tetrahedrons – one connected to tetrahedra from the triangular layer from above and from below, others are not connected to anything along the c axis. The moments from the 'bound' tetrahedra form collinear ferrimagnetic chains along the c axis; in the chains, Co^{3+} and Co^{2+} alternately alternate with antiparallel moments, the resulting moment of the subsystem $M_2 \neq 0$. The moments of the 'unbound' tetrahedra form a 120° structure with a total moment $M_1 = 0$, which corresponds to a degenerate state with two possible variants of the magnetic order: $k = 0$ or $k = (1/3, 1/3)$. As a result, at low temperatures there is a three-dimensional antiferromagnetic order (but T_N is not established). As the temperature rises, a gradual reorientation of the spins from the sublattice with 120° order takes place. The nature of neutron scattering is described with an allowance for two exchange integrals: J_1 – in the layer, J_2 – between the layers [347].

In the development of this direction, a compound was synthesized with the structure of the swedenborgite mineral $Y_{0.5}Ca_{0.5}BaCo_4O_7$, the space group $P6_3mc$ [348]. In this connection, all positions in the kagome layer are occupied by Co^{2+}, as shown in the upper panel of Fig. 7.10. The position ratio for the cobalt ions is the same as in $YBaCo_4O_7$, 3/4 of all cobalt is located in the kagome-layers (Co2), and 1/4 – in rarefied triangular layers (Co1). On the magnetic susceptibility $Y_{0.5}Ca_{0.5}BaCo_4O_7$, shown in the lower panel of Fig. 7.10, four anomalies are observed at 387, 281, 52 and 14 K. The large Weiss temperature $\Theta = -2200$ K corresponds to a very strong antiferromagnetic interaction.

The results of estimating the average magnetic moment of cobalt from neutron diffraction are in good agreement with the following proportion: 3/8 of all cobalt ions are Co^{2+} with spin $S = 3/2$, and 5/8 – Co^{3+} in the low-spin state $S = 0$ [349]. If ions with spin $S = 0$ are located in triangular layers, then the three-dimensional antiferromagnet becomes two-dimensional. However, in this case, 1/6 of the ions with spin $S = 0$ fall into the kagome layers. Nevertheless, the results of neutron diffraction are interpreted as the formation of a two-dimensional ground state of the kagome type on the spins $S = 3/2$ (Co^{2+}) with a single interaction between the nearest neighbours (long-range order does not exist). In the picture of neutron diffraction

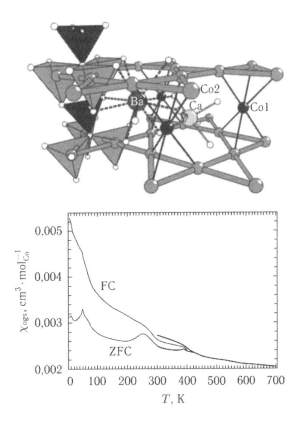

Fig. 7.10. The upper panel: the crystal structure of $Y_{0.5}Ca_{0.5}BaCo_4O_7$. The lower panel: the temperature dependence of the magnetic susceptibility $Y_{0.5}Ca_{0.5}BaCo_4O_7$ [348].

at $T = 1.2$ K there is a bright peak, which can be agreed with one of the two models of the 120°-orientation of magnetic moments in the kagome layer: a structure with $q = 0$ with a homogeneous chirality, shown in the left panel of Fig. 7.11 (the sign + corresponds to a positive chirality, when the moments in the triangle are rotated by 120° counterclockwise), due to fluctuations in the direction of the moments in the chain, they can vary in ellipses; or a structure with an alternating chirality, shown in the right panel of Fig. 7.11. In the latter case, a structure with an alternating chirality corresponds to an enlarged unit cell $\sqrt{3} \times \sqrt{3}$, ellipses represent fluctuations of the weathervane type, when there is a deviation of the moments in the hexagon.

A stable ground state can be realized by interpreting the observed neutron scattering peak as the formation of a magnetic structure with the reciprocal-lattice vector $Q = |2q_{\sqrt{3}}|$ corresponding to the structure

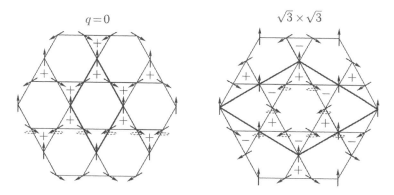

Fig. 7.11. Infinitely degenerate antiferromagnetic ground state on a kagome lattice. Ellipses show possible temperature fluctuations with zero energy. Left panel: a structure with a uniform chirality. Right panel: structure with alternating chirality [349].

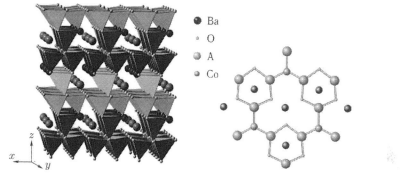

Fig. 7.12. The crystal structure of compounds with the general formula $ABaCo_4O_7$.

$\sqrt{3} \times \sqrt{3}$. Interpretation of the peak in the structure $Q = |q_0|$ will lead to the fact that the ground state can not be stabilized either by the antiferromagnetic interaction of the nearest neighbours or by the ferromagnetic interaction of the third-order neighbors. There was no peak corresponding to $Q = q_{\sqrt{3}}$, which can be explained by the absence of any magnetic interactions in the system, except for antiferromagnetic two-dimensional spin correlations between nearest neighbors.

Formation of the ground state in the series of rare-earth compounds $ABaCo_4O_7$, where A = Lu, Yb, Tm, Ho, Dy, Er, is determined by the interaction of the charge, spin and orbital degrees of freedom and is accompanied by a number of phase transitions. The crystal structure of $ABaCo_4O_7$ compounds is shown in Fig. 7.12. These substances

Table 7.1. The crystal structure parameters of $ABaCo_4O_7$ compounds (A = Lu, Yb, Tm) and the temperature of structural and magnetic transitions in them

Compound	a, Å	c, Å	T_s, K	T_m, K
$LuBaCo_4O_7$	6.263	10.225	161	47
$YbBaCo_4O_7$	6.267	10.233	178	75
$TmBaCo_4O_7$	6.276	10.240	224	105

belong to the space group $P6_3mc$ and have a hexagonal structure at room temperature [350–352]. The environment of the rare earth ion is octahedral, the Co1 and Co2 cobalt ions, belonging to different crystallographic positions, are in oxygen tetrahedra. Thus, three sublattices can be distinguished in the crystal structure: the rare-earth sublattice and two Co1 and Co2 cobalt sublattices.

The temperature dependences of the magnetic susceptibility and specific heat for $ABaCo_4O_7$ compounds (A = Lu, Yb, Tm) were studied in [350] and the results of X-ray phase analysis are presented. The data on the structure, X-ray phase analysis, temperature and field dependences of the magnetic moment for $ABaCo_4O_7$ compounds (A = Ho, Y, Tb, Dy) are presented in [351–353]. When the temperature in $ABaCo_4O_7$ (A = Lu, Yb, Tm) decreases, a structural phase transition from the hexagonal structure to orthorhombic takes place at temperatures T_s = 165, 180 and 230 K, respectively [350], the parameters of the compounds are listed in Table 7.1. X-ray diffraction analysis showed that at low temperatures the orthorhombic system has the symmetry $Cmc2_1$. The change in the volume of a unit cell during a phase transition, shown in the lower panels of Fig. 7.13, indicates that this is a first-order transition at T_s.

The temperature dependences of the resistance $ABaCo_4O_7$ (A = Lu, Yb, Tm) shown in Fig. 7.14, one can clearly see the anomaly in the form of a break associated with the structural transition at T_s. The compounds are semiconductors at room temperature. The resistivity values lie in the region ρ = 1–10 Ω cm, and the activation energies are E_a = 0.10–0.15 eV. At the structural transition the charge ordering takes place and, as a result of which, the cobalt ions Co^{2+} and Co^{3+} occupy fixed positions.

Figure 7.15 shows the temperature dependences of the inverse magnetic susceptibility $\chi^{-1}(T)$ for $ABaCo_4O_7$, measured in the field B = 0.1 T. The anomaly associated with the structural transition is best seen on the curve $\chi^{-1}(T)$ $YbBaCo_4O_7$. On the curves for the

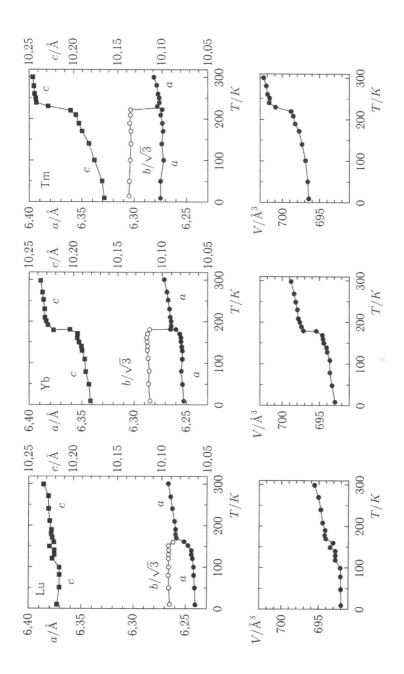

Fig. 7.13. Temperature dependences of lattice parameters (upper panels) and unit cell volume (lower panels) for ABaCo$_4$O$_7$ compounds, where A = Lu, Yb, Tm.

other two compounds, the structural transition is practically invisible. However, the slope of $\chi^{-1}(T)$ varies below the transition temperature. With a further decrease in temperature, magnetic phase transitions are observed in all compounds at temperatures $T_m = 47$, 75, and 105 K, respectively. The magnetic susceptibility of $LuBaCo_4O_7$ shows a sharp peak at $T = 47$ K, suggesting an antiferromagnetic ordering. On the $\chi^{-1}(T)$ curves measured in the FC regime, the $YbBaCo_4O_7$ and $TmBaCo_4O_7$ compounds show a deviation from the Curie–Weiss law.

For the $LuBaCo_4O_7$ compound, an effective moment of $1.73\mu_B$ was estimated. This estimate does not agree with the theoretical values for the high-spin state ($8.3\mu_B$) and the low-spin state ($6.8\mu_B$) of the Co ions). Probably, the value of the effective magnetic moment

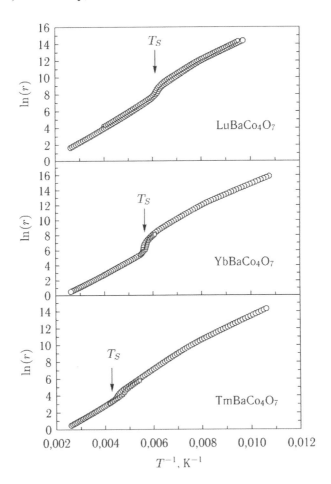

Fig. 7.14. The temperature dependences of the resistivity of $ABaCo_4O_7$, A = Lu, Yb, Tm [350].

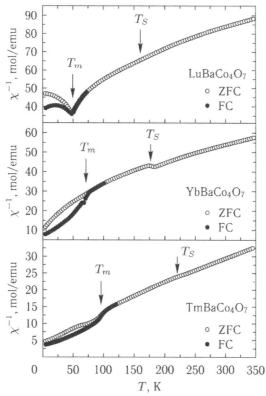

Fig. 7.15. The temperature dependences of the inverse magnetic susceptibility $ABaCo_4O_7$, A = Lu, Yb, Tm [350].

is strongly reduced because of the frustration of the magnetic interaction in the layers.

Figure 7.16 shows the temperature dependence of specific heat $C_p(T)$ of the investigated group of compounds. Each of the dependences has an anomaly in the form of a peak at a temperature close to T_S for a given composition. The anomaly in the $YbBaCo_4O_7$ sample is brighter, which may be due to the high homogeneity of the sample. At room temperature, the specific heat of all compounds reach a theoretical limit of $3Rn$ ($n = 13$). In the samples of $YbBaCo_4O_7$ and $TmBaCo_4O_7$, the anomalies at the specific heat are not visible at the temperatures of the magnetic transition. In the $LuBaCo_4O_7$ sample, the temperature dependence of the specific heat small kinks are also observed at $T = 106$ K and 45 K. In the region $45 < T < 106$ K, the temperature dependence of specific heat is linear. The break at $T = 45$ K is probably caused by magnetic ordering. The nature of the anomaly at 106 K is not clear. The temperature dependences

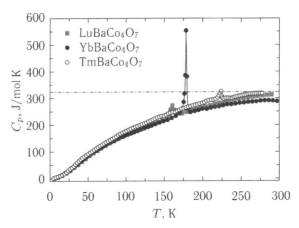

Fig. 7.16. Temperature dependences of specific heat of $ABaCo_4O_7$ compounds, where A = Lu, Yb, Tm. The dashed-dotted line shows the Dulong–Petit limit $3Rn = 324.1$ J/mol K [354].

of the heat capacity of the extended series of compounds $ABaCo_4O_7$ (A = Dy, Ho, Er, Tm, Yb, Lu) and the temperature dependence of the structural and magnetic transformations in them on the size of the rare-earth ion have been studied in Ref. [354].

Related to $YBaCo_4O_7$ is the $CaBaCo_4O_7$ compound [355–357]. Structural type 114 is close to the structure of magnetite Fe_3O_4, in which layers of octahedra with the kagome structure alternate with layers of tetrahedra with a triangular pattern, and this similarity supports interest in 114 in terms of searching for multiferroelectric behaviour. The replacement of the Y^{3+} ion by Ca^{2+} leads to the Co^{2+}: Co^{3+} ratio of 1:1 in $CaBaCo_4O_7$. Also, a significant difference in the ionic radii of yttrium and calcium leads to the fact that $CaBaCo_4O_7$ crystallizes into the orthorhombic space group $Pbn2$ close to the $YBaCo_4O_7$ structure, but with more pronounced distortions [355]. In the structure of $CaBaCo_4O_7$, there are 4 non-equivalent positions for cobalt, and a significant structural difference is the strong slope of the CoO_4 tetrahedra forming the kagome layers, as shown in Fig. 7.17. At a temperature of $T_C = 64$ K, anomalies in the dielectric constant in the form of a peak and the appearance of polarization on the $CaBaCo_4O_7$ single crystal have been observed [356, 357], as shown in Fig. 7.18. Thus, this compound is an improper multiferroic. The magnitude of the resulting polarization at 10 K is $\Delta P = 17$ mC/m² , which is a record among the improper multiferroics, it is five times higher than the value in $GdMn_2O_5$ ($\Delta P = 3600$ μC/m²) [358]. Unlike many other ferroelectrics, the structure of $CaBaCo_4O_7$

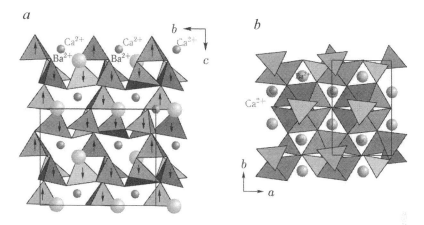

Fig. 7.17. The crystal structure of $CaBaCo_4O_7$ [355].

is non-centrosymmetric with the c polar axis. Usually at room temperature improper multiferroics possess, at room temperature, a symmetry group with a centre of inversion, which does not allow the appearance of polarization. At T_C, the magnetic order is established, which in oxides can be accompanied by a symmetry breaking and the appearance of polarization. In $CaBaCo_4O_7$, a giant polarization change is observed in magnetic fields $P(9\ T)–P(0\ T) = 8\ mC/m^2$, which is most pronounced at T_C, since magnetostriction is the basis of the strong magnetoelectric interaction during the phase transition.

In [355], the magnetic properties of $CaBaCo_4O_7$ were studied and it was established that this compound is a ferrimagnet with $T_C = 64$ K. At the same time, at high temperatures the magnetic susceptibility (the upper panel of Fig. 7.19) follows the Curie–Weiss law with a large negative Weiss temperature $\Theta = -1720$ K, similarly to $Y_{0.5}Ca_{0.5}BaCo_4O_7$. The field dependence of the magnetic moment at $T < T_C$ shows a metamagnetic transition in fields $B \sim 1$ T, and a hysteresis loop with a very high coercive force is observed (the lower panel of Fig. 7.19).

7.1.3. Langasites

To date, more than one hundred metal oxide compounds of the langasite family are known. The name 'langasite' is an abbreviation for the compound $La_3Ga_5SiO_{14}$ [359]. These materials quickly attracted considerable interest, because they possess piezoelectric properties with a high electromechanical coupling constant and

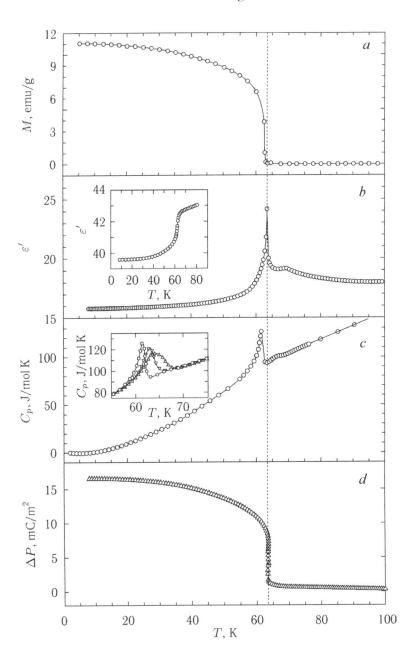

Fig. 7.18. The temperature dependences of the magnetization (*a*), the permittivity coefficient (*b*), the specific heat (*c*), and the polarization (*d*) of CaBaCo$_4$O$_7$ single crystals [357].

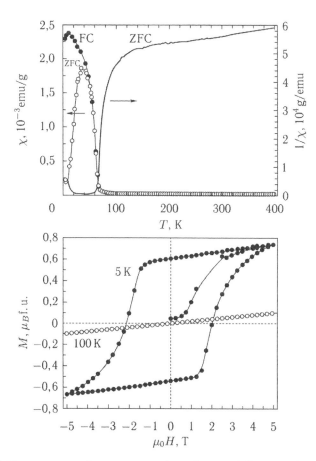

Fig. 7.19. The upper panel: temperature dependences of direct and inverse magnetic susceptibility CaBaCo$_4$O$_7$, $B = 0.3$ T. The lower panel: the field dependences of the magnetization of CaBaCo$_4$O$_7$, $T = 5$ and 100 K [355].

a lower impedance than quartz, LiNbO$_3$ or LiTaO$_3$. Many of the langasites can be grown in the form of large single crystals by the Czochralski method, and are currently used as surface acoustic wave filters in telecommunication devices and high-temperature sensors [360–362]. Langasites with magnetic ions are also an attractive class of materials in terms of studying quantum ground states, having a distorted kagome lattice.

Most of the compounds of the langasite family have the structure of Ca$_3$Ga$_4$Ge$_2$O$_{14}$, shown in Fig. 7.20: the trigonal lattice with the space symmetry group $P321$ [363, 364]. There are two layers in the langasite structure: the $z = 0$ layer formed by large oxygen polyhedra, and the $z = 1/2$ layer with two types of oxygen tetrahedra. In each layer there are two types of polyhedra, only four positions for

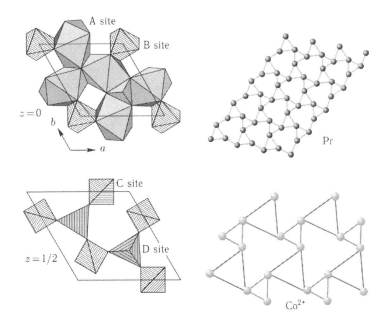

Fig. 7.20. The left panel: the crystal structure of the substances of the langasites [364]. The right panel: distorted kagome grid in the $z = 0$ layer occupied by $4f$ metals ions in position A (A = Pr for example) [364], and a grid of triangles in the $z = 1/2$ layer occupied by $3d$ metal ions in position D (D = Co for example).

cations: a Thompson cube and an octahedron in the $z = 0$ layer, large and small tetrahedra in the $z = 1/2$ layer. These sites for cations can be occupied by four different ions, or the same cation can be located in polyhedra of different types. The general formula is written as $A_3BC_3D_2O_{14}$, where the letters A, B, C and D denote four different positions. In order to organize a frustrated exchange interaction in langasites, one should create substances in which magnetic ions are placed in structural positions A and/or D, which form kagomes and triangular sublattices, respectively, as shown in the right panel of Fig. 7.20.

The presence of four positions for cations suggests a large number of compounds in this family, in particular, the following variants are known. The layer $z = 0$: AO_8 (cube) can be occupied by ions La^{3+}, Nd^{3+}, Pr^{3+} or Ba^{2+}, Sr^{2+}, Ca^{2+}, Pb^{2+}; BO_6 (octahedron) – by the ions Ga^{3+}, Ta^{5+}, Nb^{5+}, Sb^{5+} or Te^{6+}. The $z = 1/2$: CO_4 (large tetrahedron) layer can be occupied by the ions Ga^{3+}, Al^{3+}, Fe^{3+} or Zn^{2+}, Mn^{2+}, Co^{2+}; DO_4 (small tetrahedron) – by the ions Ge^{4+}, Si^{4+} or P^{5+}, As^{5+}, V^{5+}.

A number of studies have been carried out on the compounds $Pr_3Ga_5SiO_{14}$ and $Nd_3Ga_5SiO_{14}$, where the rare-earth ion occupies

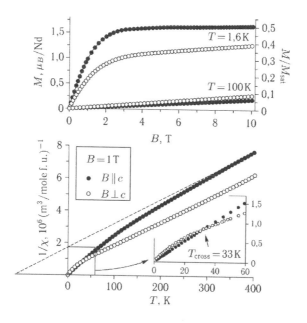

Fig. 7.21. The field dependence of the magnetization and the temperature dependence of the inverse magnetic susceptibility of the Nd$_3$Ga$_5$SiO$_{14}$ single crystal [366].

position A in the structure [364–371]. In these compounds, the magnetic ions Nd^{3+} ($J = 9/2$) or Pr^{3+} ($J = 4$) are located at the nodes of the distorted kagome grid, as shown in the right panel of Fig. 7.20. It was found that the magnetic order in Nd$_3$Ga$_5$SiO$_{14}$ and Pr$_3$Ga$_5$SiO$_{14}$ is not established up to 40 mK, as can be seen from the temperature dependences of the magnetic susceptibility shown in Fig. 7.21 and 7.22 [366]. The values of the paramagnetic Weiss temperature are $\Theta = -52$ K in Nd$_3$Ga$_5$SiO$_{14}$ and $\Theta = -2.3$ K in Pr$_3$Ga$_5$SiO$_{14}$. The minus sign corresponds to the AFM interaction between the moments, and the behaviour of $\chi^{-1}(T)$ follows the Curie–Weiss law in the interval 50–300 K. A crossover of the magnetic susceptibility along the directions of the external field $B \parallel c$ and $B \perp c$ is observed $T_{\mathrm{cross}} = 33$ K for Nd$_3$Ga$_5$SiO$_{14}$ and $T_{\mathrm{cross}} = 127$ K for Pr$_3$Ga$_5$SiO$_{14}$. A change in the character of the temperature dependence indicates that at high temperatures in both compounds there is an anisotropy of the easy-plane type and the moments lie in the plane ab. As the temperature is lowered, the crystal field changes, and the c-axis becomes the easy axis of magnetization.

The absence of long-range order allowed the authors of [364–371] to nominate rare-earth langasites as candidates for a. ground

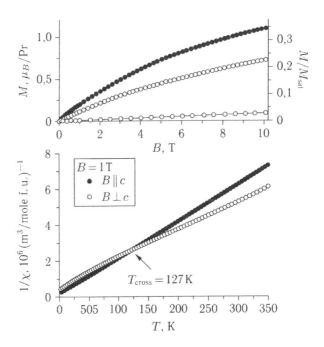

Fig. 7.22. The field dependence of the magnetization and the temperature dependence of the reciprocal magnetic susceptibility of the $Pr_3Ga_5SiO_{14}$ single crystal [366].

state of the spin-liquid type with short-range AFM correlations between nearest neighbors. This is confirmed by experiments on neutron scattering, in which only diffuse scattering is observed, and there are no peaks corresponding to magnetic ordering. Such a scattering pattern is preserved up to temperatures of the order of mK in $Nd_3Ga_5SiO_{14}$.

There are subtle differences in the behaviour of compounds with neodymium and praseodymium at low temperatures. In $Nd_3Ga_5SiO_{14}$, a spin liquid with strong AFM bonds is realized in the ground state, since frustrations of the exchange interaction do not allow the formation long-range order, despite the substantial value of the paramagnetic Weiss temperature (~ -60 K). Under the influence of an external field of 0.6 T, a field-induced transition to an ordered state is observed at $T_C = 0.28$ K [368]. In this case, the diffuse scattering of neutrons becomes low-intensity, and the intensity of the magnetic Bragg peaks grows, and the entire diffraction pattern in the field resembles a grid of ferromagnetically oriented moments on Nd^{3+} ions.

In $Pr_3Ga_5SiO_{14}$, the interaction between the neighbouring moments is very weak (~ -2 K) and the system behaves like a cooperative paramagnet. Diffuse neutron scattering in $Pr_3Ga_5SiO_{14}$ is not observed, and under the action of the external magnetic field $B \perp ab$ a gap appears in the spectrum of magnetic excitations [370]. In addition, the temperature dependence of specific heat without a magnetic field exhibits a Schottky-type anomaly at $T = 6.7$ K, due to the splitting of the Pr^{3+} ion levels by the crystal field. The anomaly is suppressed by an external field of 9 T, and the calculated magnetic contribution to the heat capacity has the form $C_{mag} = AT^{\alpha}$, where the exponent α = 1.98 corresponds to two-dimensional spin excitations (magnons) in the zero field [365, 370–371]. Thus, in $Pr_3Ga_5SiO_{14}$ without a magnetic field, as $T \to 0$, over a degenerate ground state, there exists a gas of two-dimensional magnons, and under the action of an external field $B = 9$ T, a short-range correlation regime is formed: nanoislands (25–30 Å) of an ordered phase appear inside which there is a gap between the ground and excited states [370].

Synthesis and structural studies of the new beryllium subgroup of the langasite family with the general formula $A_3Ga_3Ge_2BeO_{14}$ (A = La, Nd, Pr, Sm) are described in [372]. From the point of view of the magnetic subsystem, these compounds are analogous to the above-described $A_3Ga_5SiO_{14}$ compounds, since they contain magnetic $4f$ ions in A position. Field dependences of the magnetic moment and the temperature dependences of the magnetic susceptibility of the $Nd_3Ga_3Ge_2BeO_{14}$ and $Pr_3Ga_2Ge_2BeO_{14}$ compounds are shown in Figs. 7.23 and 7.24, respectively [373]. It can be seen that the $\chi(T)$ dependences do not exhibit anomalies that could be attributed to the formation of long-range order. Dependences of $M(B)$ at low temperatures for $Nd_3Ga_3Ge_2BeO_{14}$ quickly reach saturation with a sufficiently small value of the magnetic moment of $1.25\mu_B/Nd^{3+}$, which, in analogy with the $Nd_3Ga_5SiO_{14}$ discussed above, can be caused by the anisotropy of the Nd^{3+} ion. The field dependences of the magnetization for $Pr_3Ga_3Ge_2BeO_{14}$ do not reach saturation, however, the value of the magnetic moment in the field 7T is also small, $0.45\mu_B/Pr^{3+}$.

It was found that the magnetic order in $A_3Ga_3Ge_2BeO_{14}$ (A = Nd, Pr, Sm) is not established up to 500 mK. The $\chi^{-1}(T)$ dependence deviates from the Curie–Weiss law at $T < 50$ K in the Pr composition, at $T < 100$ K in the Nd composition, and in the Sm composition does not obey this law in the entire temperature range. This behaviour is attributed to the van Vleck paramagnetism for rare-earth ions, caused by a change in the population of the sublevels of the main multiplet upon cooling [373]. The values of the Weiss paramagnetic

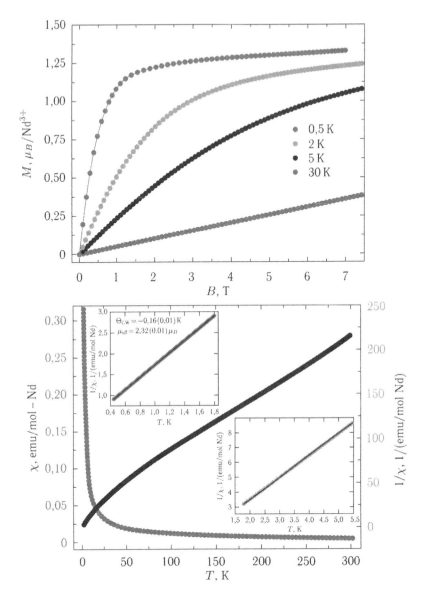

Fig. 7.23. The field dependence of the magnetization and the temperature dependence of the magnetic susceptibility of $Nd_3Ga_3Ge_2BeO_{14}$ [373].

temperature, estimated from the high-temperature region, are $\Theta = -38$ K in $Nd_3Ga_3Ge_2BeO_{14}$ and $\Theta = -40$ K in $Pr_3Ga_3Ge_2BeO_{14}$.

The temperature dependence of specific heat of beryllium langasites was also studied in [373]. Anomalies corresponding to the formation of long-range magnetic order were not detected up to a temperature of 350 mK, which allows the authors to propose that

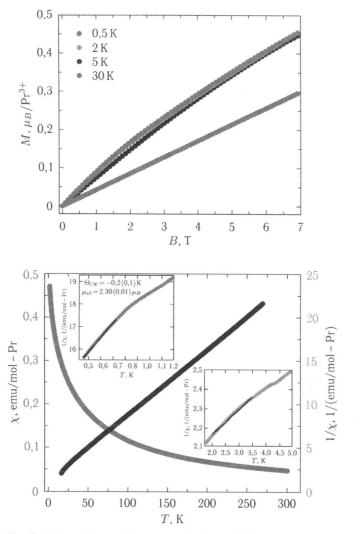

Fig. 7.24. The field dependence of the magnetization and the temperature dependence of the magnetic susceptibility of $Pr_3Ga_3Ge_2BeO_{14}$ [373].

the ground spin state is the spin-liquid state, however, for the final derivation, experiments are necessary on neutron scattering and muon depolarization in $A_3Ga_3Ge_2BeO_{14}$ (A = Nd , Pr, Sm).

Much attention is paid to the properties of langasites containing the Fe^{3+} ion at the D position of the structure (large tetrahedra in the $z = 1/2$ layer) [249, 250, 364, 374–377]. In this layer, 3d-ions form a two-dimensional grid composed of large and small triangles, as shown in the left panel of Fig. 7.20. All the investigated iron langasites with the general formula $A_3BFe_3D_2O_{14}$ (A = Ba, Sr, Ca,

B = Ta, Nb, Sb, D = Ge, Si) have close Néel temperatures $T_N = 26$–36 K and Weiss temperatures $\Theta = 100$–200 K. The Néel temperature essentially depends on the size of the diamagnetic ion located in the inactive layer $z = 0$ in position B, that is, the long-range magnetic order is established due to the interlayer exchange.

The Fe-based $Ba_3NbFe_3Si_2O_{14}$ is an improper multiferroic, i.e., below the Néel temperature $T_N = 27$ K, the polarization vector **P** and the spiral magnetic structure on the basis of triangular prisms of Fe^{3+} ions are simultaneously observed in it, as shown in Fig. 7.25 [364]. The appearance of tubes elongated along the c axis became unexpected for a layered structure where a strong magnetic interaction is assumed in the ab (J_1 and J_2) plane and much weaker along the c axis. It turned out that along the c axis there are diagonal exchange interactions between the layers J_3, J_4 and J_5, which lead to the formation of a magnetic spiral with a period of τ ~ 1/7, shown in Fig. 7.25, while in the triangles in the layer ab the magnetic moments are unfolded at an angle of 120°.

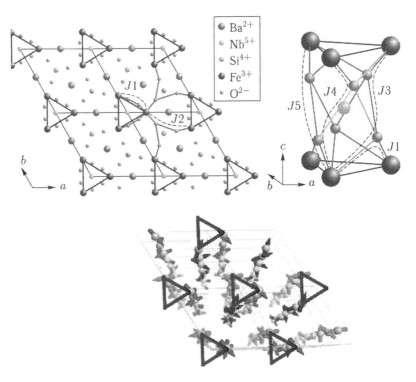

Fig. 7.25. The magnetic structure of is $Ba_3NbFe_3Si_2O_{14}$ [364]. Top panel: the ways of exchange interaction in the layer and between layers. Bottom panel: spiral structure at $T < T_N$.

Table 7.2. Parameters of the exchange interaction in $Ba_3NbFe_3Si_2O_{14}$ as a function of the Hubbard energy U [376]

	$U = 0$ eV	$U = 3$ eV	$U = 5$ eV
J_1, K	40.1 (1.00)	25.7 (1.00)	20.1 (1.00)
J_2, K	10.5 (0.26)	6.3 (0.25)	4.4 (0.22)
J_3, K	2.7 (0.07)	1.3 (0.05)	0.9 (0.04)
J_s, K	19.5 (0.49)	11.2 (0.44)	8.0 (0.40)

Calculations from the first principles in the GGA + U technique, performed in [376], give estimates of the exchange integrals given in Table 7.2. It is seen that all the J_i have an antiferromagnetic character. As expected, the exchange inside the layer J_1 is maximal, the direct interlayer exchange J_5 is minimal, but the sum of the diagonal interlayer integrals $J_5 = J_3 + J_4$ is only half that of J_1. The resulting values of J_i depend on the parameter of the model U, however, when the obtained Weiss temperature is compared with its experimental value $\Theta = -190$ K, the set of exchange integrals for $U = 5$ eV is the closest to the experiment.

The temperature dependence of permittivity ε, polarization, and also the polarization loop $P(E)$, observed below T_N in a $Ba_3NbFe_3Si_2O_{14}$ single crystal, are shown in the upper panel of Fig. 7.26 *a, b*. The appearance of polarization in a zero magnetic field is forbidden for the symmetry group $P321$, which has a centre of inversion. For the appearance of polarization, the substance must undergo a structural phase transition to the low-symmetric phase, for example into monoclinic $C2$, which will be due to the loss of the third-order symmetry axis. However, the $P321$ and $C2$ groups are virtually indistinguishable for conventional X-ray diffraction analysis, especially if small monoclinic distortions are present in the starting compound. Until now, the X-ray spectra at low temperatures have not been completely attributed to the symmetry of $C2$, although the anomaly of the lattice parameter a for T_N is noted [364]. Nevertheless, this symmetry group is the basis for the analysis of neutron diffraction data in $Ba_3NbFe_3Si_2O_{14}$ [377]. The results of this description are shown in the lower part of Fig. 7.26, model calculations are carried out using an additional temperature for the formation of the helicoid of polarization $T_p \neq T_N$.

Thus, in the langasites with Fe^{3+}, the ground state is the long-range AFM order. The frustrations of the exchange interaction,

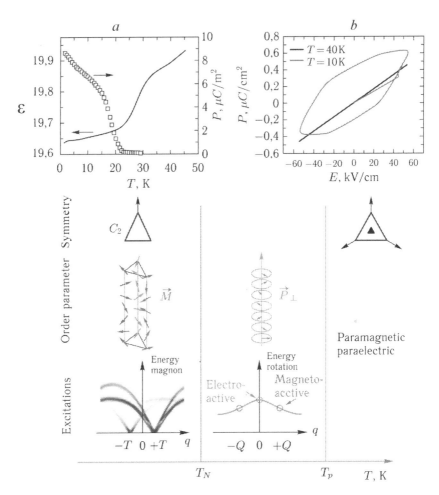

Fig. 7.26. The upper panel: the temperature (*a*) and field (*b*) polarization dependences of $Ba_3NbFe_3Si_2O_{14}$ [375]. Bottom panel: the temperature evolution of the electrical and magnetic phases. $T > T_p$ – paramagnetic and paraelectric state, $T_N < T < T_p$ – helicoidal polarization, $T < T_N$ – magnetic helicoid [377].

imposed by a triangular motif in the layer, are overcome by interlayer diagonal interactions.

7.1.4. Dugganites

Compounds from the langasite subclass containing Te^{6+} ions (dugganites) and magnetic ions of cobalt and manganese also turned out to be 'exotic' magnets. Actually, the mineral dugganite, $Pb_3TeZn_3As_2O_{14}$, is not magnetic, but its manganese and cobalt structural analogues contain magnetoactive 3*d* ions at position D and are antiferromagnetically ordered at low temperatures [378–386]. The crystal structure of the dugganites has a monoclinic symmetry group $P121$, which is the second derived subgroup of the trigonal space group $P321$, which is characteristic of the langasites. In the crystalline structure of cobalt dugganites ($Pb_3TeCo_3V_2O_{14}$, $Pb_3TeCo_3P_2O_{14}$ and $Pb_3TeCo_3As_2O_{14}$), large CoO_4 tetrahedra and small tetrahedrons of PO_4 (VO_4) form slightly corrugated layers with a triangular motif in the arrangement of magnetoactive Co^{2+} ions, as shown in the upper panel of Fig. 7.27. The specific feature of the crystal lattice is the large number of monoclinic distortions of oxygen tetrahedra, resulting in 12 non-equivalent positions in the structure for the cobalt ion, which significantly complicates structural analysis and first-principles calculations. Magnetoactive layers alternate with non-magnetic layers formed from the decahedrons PbO_8 and octahedra TeO_6. In this case, lead ions Pb^{2+} are organized in a zigzag manner inside non-magnetic layers. Cobalt ions in magnetoactive layers are linked either by means of two oxygen atoms belonging to the tetrahedra VO_4 (large cobalt triangles) or by two oxygen atoms belonging to TeO_6 octahedra (small cobalt triangles). A scheme with two kinds of triangles was shown in Fig. 7.20. Interlayer exchanges are realized through TeO_6 octahedra in non-magnetic layers.

The magnetic and structural properties of the compound $Pb_3TeCo_3V_2O_{14}$ have become the subject of active research [379–384]. As the temperature is lowered, the ground antiferromagnetically ordered state in the $Pb_3TeCo_3V_2O_{14}$ compound is achieved by forming a regime of short-range correlations at $T^* \sim 10.5$ K and successive phase transitions of the second order at $T_{N1} \sim 9$ K and of the first order $T_{N2} \sim 6$ K. The magnetic structure with the vector $\mathbf{q}_1 = (1/2, 0, -1/2)$ is formed in the temperature range $T_{N2} < T < T_{N1}$, which in the interval $T < T_{N2}$ is rearranged into a structure with the vector $\mathbf{q}_2 = (1/2, 1/2, -1/2)$ [382]. The external magnetic field destroys the ordered magnetic structure in the temperature range $T < T_{N2}$ and affects the magnetic order in the range $T_{N2} < T < T_{N1}$, which leads to a complex magnetic phase diagram of $Pb_3TeCo_3V_2O_{14}$ shown in the lower panel of Fig. 7.27 [383]. Phases I and II correspond to different antiferromagnetic ordered states. Phase III is a state with an oblique antiferromagnetic order. Phase IV is a paramagnetic state.

The phase transitions at T_{N1} and T_{N2} are accompanied by anomalies of the magnetic susceptibility and heat capacity as shown in Fig. 7.28, as well as the anomaly of the dielectric constant.

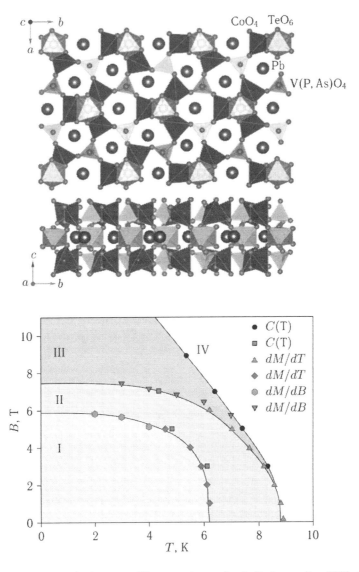

Fig. 7.27. Upper panel: the crystalline structure of cobalt dugganites [383, 384]. Lower panel: the magnetic phase diagram of the $Pb_3TeCo_3V_2O_{14}$ compound [383].

Calculations of exchange integrals from the first principles showed that, despite the layered crystal structure, the magnetic subsystem in $Pb_3TeCo_3V_2O_{14}$ consists of quasi-one dimensional triangular tubes elongated along the c axis, as shown in the left panel of Fig. 7.29 [383]. The main paths of intralayer and interlayer exchange interactions are shown by solid and dotted arcs. In contrast to the

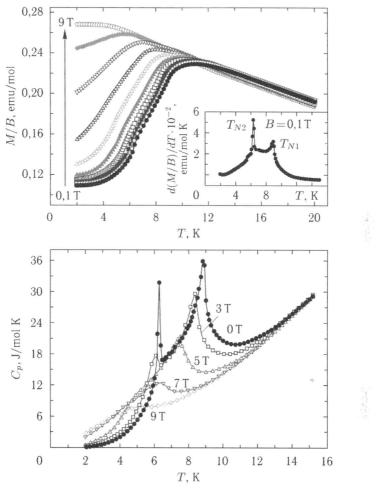

Fig. 7.28. Temperature dependences of the magnetic susceptibility (top) and heat capacity (bottom) near the double phase transition in $Pb_3TeCo_3V2O_{14}$ in various magnetic fields. On the inset – the derivative of the magnetic susceptibility [383].

previously considered iron langasites, where a 120°N noncollinear structure was formed in the layer *ab* as a result of the AFM exchange on the triangles, and the moments were in the layer, the magnetic exchange inside the small cobalt triangles in $Pb_3TeCo_3V_2O_{14}$ is ferromagnetic, $J_2 = -1.4$ K , and it is much larger than the exchange inside large cobalt triangles, $J_1 = -0.2$ K. As a result, the main triangular cluster has a collinear magnetic structure, with moments directed perpendicular to the layers. Thus, the system should be regarded as a network of weakly connected triangles in the layer, since a strong exchange interaction acts along the *c* axis between

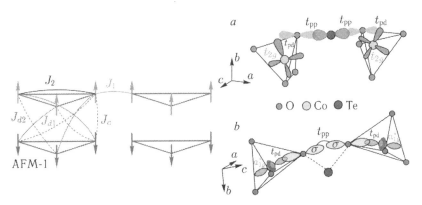

Fig. 7.29. Left panel: magnetic configuration and main paths of exchange interactions in $Pb_3TeCo_3V_2O_{14}$. Right panel: a schematic representation of supersuper-exchange paths for diagonal interlayer exchange interactions J_{d1} and J_{d2}.

triangles lying in neighboring Co–V planes. This connection is arranged in a very non-trivial way. The exchange integral along the c axis is small, amounting to $J_c = 2.7$ K; however, the diagonal exchanges indicated in Fig. 7.29 as J_{d1} and J_{d2}. Exact calculations of specific values of the diagonal exchanges are hampered by the large value of the unit cell ($Z = 6$), but an estimate of the average diagonal exchange shows that it is antiferromagnetic $J_{d1} + J_{d2} = 4.3$ K and is the largest in the system studied. Together with ferromagnetic J_c and J_2, this exchange removes frustration and leads to a general antiferromagnetic collinear order for Co ions along the c axis in agreement with the experimental data.

An important result of the calculations was the fact that the interlayer exchange J_{d1} and J_{d2} for Co–Co contacts located at distances of ~ 6.29 Å turned out to be comparable with the exchange of J_2 within small cobalt triangles, where the distance between interacting Co–Co ions is almost two times less than ~ 3.52 Å. Such a strong exchange interaction at such distances in insulators can only be obtained by supersuper-exchange. In the structure there are two types of atoms located along the interlayer diagonals: Te and O, which are organized in two different ways. Along one diagonal Te and two O ions form a connection with an angle close to 180° (Fig. 7.29a), while along the other diagonal the coupling angle is close to 90° (Fig. 7.29b). Thus, we can assume two possible supersuper exchange paths along these two diagonals. In the case of a 180° connection, super-exchange occurs between two t_{2g} orbitals through two oxygen $2p$ orbitals and one p-orbital Te (Fig. 7.29a). In the case of the 90° connection, we can assume a more elegant path: since the

t_{2g} orbitals for the tetrahedrally coordinated Co^{2+} ion are half-filled (one electron per orbital with parallel spins), we can construct the orbit $a_{1g} = (d_{xy} + d_{yz} + d_{zx})/\sqrt{3}$, which strongly overlaps with the oxygen orbital $p_\sigma = (p_x + p_y + p_z)/\sqrt{3}$ [387]. The orbital p_σ on two different oxygen ions forming the 90° O–Te–O bond are directed at each other in such a way that the super-exchange process includes two pd paths (from Co a_{1g} to O p_σ) and one pp path (between the two orbital p_σ) for virtual electron hopping, as shown in Fig. 7.29 *b*. Thus, taking into account the strong interlayer exchanges between small cobalt triangles, the $Pb_3TeCo_3V_2O_{14}$ compound can be regarded as a quasi-one dimensional system of weakly coupled triangular tubes.

Unlike the compound containing vanadium $Pb_3TeCo_3V_2O_{14}$, where the magnetically ordered state is achieved by two successive phase transitions, compounds containing phosphorus and arsenic are ordered antiferromagnetically at a single Néel temperature equal to $T_N = 12.8$ and 16.7 K for $Pb_3TeCo_3P_2O_{14}$ and $Pb_3TeCo_3As_2O_{14}$, respectively.

Data on the spin dynamics obtained by the ESR method on powder samples of $Pb_3TeCo_3V_2O_{14}$, $Pb_3TeCo_3P_2O_{14}$, and $Pb_3TeCo_3As_2O_{14}$ with a temperature variation are shown in Fig. 7.30. The system with vanadium reveals a weak anisotropic signal characteristic of Co^{2+} ions in tetrahedral oxygen coordination (Fig. 7.30*a*). The ground state of Co^{2+} ions in a tetrahedral crystalline field is fourfold degenerate, in the first approximation e_g the orbital is completely filled, and t_{2g} is half-filled. In this case, the spin magnetic moment of the Co^{2+} ion is $S = 3/2$ in the absence of the orbital angular momentum. In the second approximation of perturbation theory, the Coulomb repulsion and the exchange interaction (the multiplets effect) mix a certain number of e_g electrons with t_{2g} electrons and partially restore the orbital angular momentum, which leads to deviation of the g-factor from a purely spin one close to $g = 2$. Such a mixing depends on on the splitting of the *d*-shell by the ligand field, and on the spin-orbit interaction. It does not split the ground state, but lowers its energy. For the spin system $S = 3/2$ (high-spin Co^{2+}) zero field splits the energy levels into two doublets $|\pm 1/2\rangle$ and $|\pm 3/2\rangle$. Thus, the observed absorption spectrum can be ascribed to transitions between Kramers doublets with an effective spin of 3/2 [41, 388]. For quantitative estimates, the experimental ESR spectrum was approximated by the sum of two Lorentzians corresponding to the main (longitudinal and transverse) components of the anisotropic g-tensor:

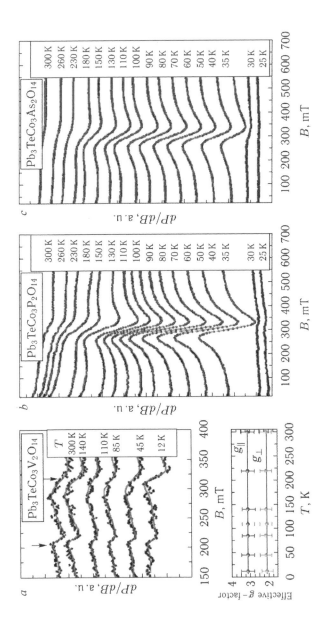

Fig. 7.30. Evolution of the ESR spectra of powder samples of $Pb_3TeCo_3V_2O_{14\,(a)}$, $Pb_3TeCo_3P_2O_{14\,(b)}$, and $Pb_3TeCo_3As_2O_{14\,(c)}$ at a temperature variation: symbols are experimental data, solid lines are the result of the approximation of the absorption line by one or two Lorentzians, as described in the text. The lower part of the panel (a) shows the temperature dependences of the principal components of the anisotropic

$$\frac{dP}{dB} \propto \frac{d}{dB}\left[\frac{\Delta B}{(B - B_r)^2 + \Delta B^2}\right], \tag{7.2}$$

where B_r is the resonant field, and ΔB is the width of the absorption line. The result of the approximation is shown by solid lines in Fig. 7.30 *a*. The obtained values of $g_\parallel \approx 3.10$ and $g_\perp \approx 2.13$ are in agreement with the experimental values observed for other compounds with Co^{2+} in tetrahedral coordination [389–391]. Using the average effective g-factor $g = 1/3\ (g_\parallel + 2g_v) \approx 2.45$, we can estimate the effective magnetic moment $\mu_{theor} = 8.2\mu_B$/f.u., which is in good agreement with the estimate from the temperature dependence of the magnetic susceptibility $d\Delta B/dT$. An estimate of the exchange anisotropy, according to the expression $J_\perp/J_\parallel = g_\perp/g_\parallel$ [392], gives the value $J_\perp/J_\parallel \sim 0.7$, which indicates a moderate magnetic anisotropy in the $Pb_3TeCo_3V_2O_{14}$ system. When the temperature is lowered, the value of the g-factor remains practically unchanged (the lower panel in Fig. 7.30a), and the line width decreases noticeably, which corresponds to a weakening of the role of the spin-lattice relaxation.

The character of the absorption spectra of ESR and spin dynamics in dugganites with P and As ions is practically identical. A single wide line corresponding to the signal from the Co^{2+} magnetic ions (Fig. 7.30 *b*, *c*) is observed in the entire temperatures. Because of the substantial (at least triple) broadening of the absorption line in comparison with the data for the vanadium sample, the anisotropy effects are masked, but the shape of the absorption line remains asymmetric. For quantitative estimates of the principal ESR parameters, the experimental spectra were approximated by a single function of the Dyson type (4.1). The results of the approximation are shown by solid lines in Fig. 7.30 *b*, *c*. The asymmetry parameter takes the value $\alpha \sim 0.4$, indicating a significant contribution of the dispersion to the absorption curve.

The temperature dependences of the effective g-factor and the absorption line width for $Pb_3TeCo_3P_2O_{14}$ and $Pb_3TeCo_3As_2O_{14}$ are shown in Fig. 7.31. Analogous to the data for the dugganite with V, the g-factor for P and As samples is practically independent of temperature, and the line width decreases with decreasing temperature. The average values of the g-factors are $g = 2.19 \pm 0.02$ for $Pb_3TeCo_3P_2O_{14}$ and $g = 2.21 \pm 0.02$ for $Pb_3TeCo_3As_2O_{14}$, respectively. The line width increases linearly with $T > 50$ K. This behaviour can be related to the fast spin-

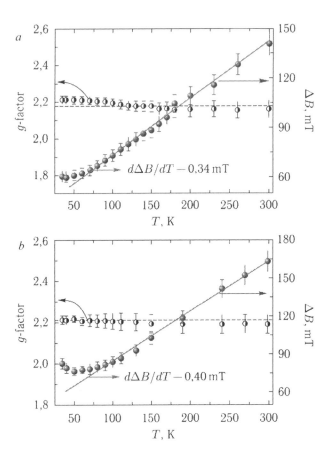

Fig. 7.31. The temperature dependences of the effective g-factor and the width of the absorption line for the $Pb_3TeCo_3P_2O_{14}$ (a) and $Pb_3TeCo_3As_2O_{14}$ (b).

lattice relaxation that is expected for concentrated exchange-coupled systems, where the resulting spin states can provide more efficient channels for relaxation processes [94].

Thus, a detailed study of the dynamic magnetic properties of three antiferromagnetic synthetic analogues of dugganite, $Pb_3TeCo_3V_2O_{14}$, $Pb_3TeCo_3P_2O_{14}$, and $Pb_3TeCo_3As_2O_{14}$, revealed the behaviour characteristic of quasi-one-dimensional systems. This behaviour was explained by first-principles calculations, which showed that, despite the layered crystal structure, the magnetic subsystem has a quasi-one-dimensional character and is a weakly coupled triangular tube.

7.2. Plateau of magnetization in the lattice of diamond chains

The triclinic crystal structure of sodium–iron phosphite $NaFe_3(HPO_3)_2(H_2O_3)_6$ has a space group $P1$ [393]. From the point of view of magnetic topology, the $NaFe_3(HPO_3)_2(H_2O_3)_6$ compound is characterized by a non-trivial magnetic lattice with two different subsystems corresponding to two crystallographically nonequivalent positions of Fe^{3+} ions (Fig. 7.32). Fe (1) ions form dimers (or alternating chains) in the structure that connect through two phosphate groups of HPO_3, while Fe (2) ions connect these chains to the common three-dimensional framework through the phosphate groups of H_2PO_3.

The temperature dependences of the *dc* and *ac* magnetic susceptibility at 0.1 T for the $NaFe_3(HPO_3)_2(H_2O_3)_6$ compound are shown in Fig. 7.33 [394]. In general, the $\chi(T)$ dependence shows Curie–Weiss behaviour with decreasing temperature, then exhibits a bending at $T \sim 10$ K, which is replaced by a sharp increase in χ_{dc} and an acute maximum in χ_{ac}, which indicates a phase transition to the ferrimagnetic state at $T_C \sim 9.5$ K. Analysis of the high-temperature part (200–300 K) $\chi(T)$ in accordance with the Curie–Weiss law satisfactorily describes the experimental data at the following values of the parameters: $C = 13.4$ emu K/mol and $\Theta = -21$ K for the $\chi_{dc}(T)$

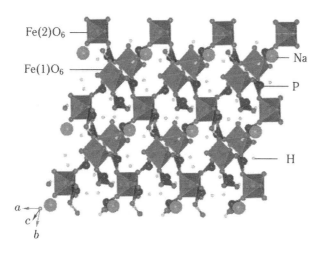

Fig. 7.32. Polyhedral form of the crystal structure of the $NaFe_3(HPO_3)_2(H_2PO_3)_6$ compound. The FeO_6 octahedra for the two non-equivalent positions of Fe^{3+} ions are shown in different shades.

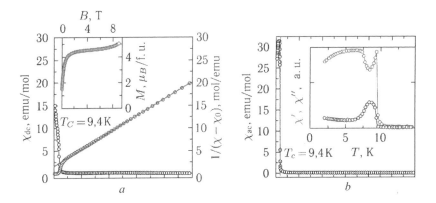

Fig. 7.33. Temperature dependences of *dc-* (*a*) and *ac-* (*b*) magnetic susceptibility of NaFe$_3$(HPO$_3$)$_2$(H$_2$PO$_3$)$_6$. On the inserts: *a*) the field dependence of the magnetization in static magnetic fields; *b*) the real χ' and imaginary χ'' parts of the magnetic susceptibility at low temperatures.

and C = 12.6 emu K/mol and Θ = -22 K for the $\chi_{ac}(T)$ dependence. The effective magnetic moment, estimated from the Curie constant $\mu_{eff} \approx 10.3\mu_B$/f.u. completely agrees with the theoretical value $\mu_{theor} \approx 10.3\mu_B$/f.u., where the value of the effective *g*-factor, obtained from the ESR spectra, was *g* = 1.99.

Thedata on the specific heat of NaFe$_3$(H$_2$PO$_3$)$_6$(HPO$_3$)$_2$ are in agreement with the data on the magnetic susceptibility and confirm the transition to a magnetically ordered state at low temperatures (Fig. 7.34). The temperature dependence of spefific heat $C_p(T)$ in a zero magnetic field exhibits an anomaly of the λ type at $T_C \sim 9.5$ K. When an external magnetic field is applied, the anomaly on $C_p(T)$ broadens and shifts to higher temperatures.

At room temperature, the specific heat reaches half the Dulong–Petit thermodynamic limit, as shown by the horizontal line in Fig. 7.34. The contribution of the lattice C_{lat}, shown by the dashed line in the inset in Fig. 7.34, is negligibly small at low temperatures, and the corresponding magnetic contribution of C_p shows an almost linear behaviour characteristic of quasi-2D ferrimagnetic systems [395].

Figure 7.35 shows the field dependences of the magnetization for NaFe$_3$(H$_2$PO$_3$)$_6$(HPO$_3$)$_2$. The magnetization curve reveals a plateau at one-third of the saturation moment $M_s/3 \sim 4.3\mu_B$ in the field interval 2–9 T and reaches saturation at $B_s \sim 27$ T at a temperature T = 2.4 K. The saturation moment $M_s \sim 13\mu_B$ turned out to be somewhat lower than the theoretically expected of the moment $\sim 15\mu_B$, which can be related both to the final temperature at which the measurement was carried out and to the effects of overheating in measurements in pulsed fields. The data on the specific heat at $B = 3$, 6, and 9 T (the insert in Fig. 7.35) is in accordance with the behaviour of the magnetization, which demonstrates the plateau in this interval of fields.

The evolution of the ESR spectra with temperature is shown in Fig. 7.36 *a*. At high temperatures, the exchange-narrowed absorption line of the Lorentzian form typical of magnetically concentrated systems is observed in the spectrum. It turned out, however, that the correct description of the line shape requires the use of the sum of two Lorentz functions. This means that in the spectra there are two different resonant modes L_1 and L_2. Taking into account the structure of the crystal structure, it is natural to assign these lines to signals from two types of Fe^{3+} ions. There is a satisfactory agreement between the experimental and theoretical relationships over the entire temperature range studied. An example of the expansion of the spectrum into two components is given in the upper part of Fig. 7.36 *a*.

The temperature dependences of the effective *g*-factor, the width of the ESR line, and the integrated intensity for the resolved spectral components are shown in Fig. 7.36 *b*. At high temperatures, two exchange-narrowed absorption lines are characterized by isotropic effective *g*-factors $g_1 = 1.99$ and $g_2 = 1.99$, typical of high-spin Fe^{3+} ($S = 5/2$) in the octahedral coordination of oxygen. As the temperature is lowered, the resonance field of the L_1 line increases noticeably, which indicates the development of short-range order correlations. At the same time, the resonance field of the L_2 line does not deviate from a constant value to a lower temperature, and the apparent shift of the resonance mode begins only in the immediate vicinity of the phase transition temperature (the upper panel in Fig. 7.36 *b*).

The different nature of the two resonant modes is most clearly manifested in the behaviour of the integrated intensity of the ESR χ_{ESR} (the lower panel in Fig. 7.36 *b*). The integrated ESR intensity χ_{ESR2} of the mode L_2 demonstrates the behaviour of the Curie–Weiss type when the temperature is lowered to ~30 K, and then passes through a maximum and decreases.

At the same time, χ_{ESR1} passes through a wide maximum near 90 K, reminiscent of the behaviour of low-dimensional spin-slit systems. Bearing in mind that Fe (1) ions form alternating spin chains connected in the structure by two HPO_3 phosphate groups, it can be assumed that the L_1 mode corresponds to the signal from

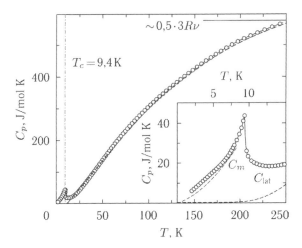

Fig. 7.34. The temperature dependence of spefific heat of $NaFe_3(HPO_3)_2(H_2O_3)_6$.

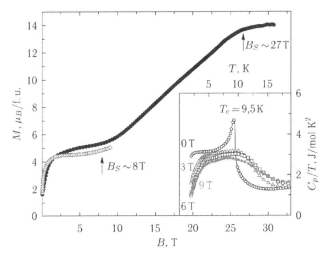

Fig. 7.35. The field dependence of the magnetization of $NaFe_3(H_2PO_3)_6(HPO_3)_2$ at $T = 2.4$ K (in pulsed fields) and at $T = 2$ K (in static fields). On the inset: the normalized specific heat at a variation of the magnetic field.

this low-dimensional magnetic sublattice, while the L_2 mode can be associated with the signal from Fe (2) ions that connect these chains to the overall 3D structure also through phosphate groups. At $T^* \sim 30$ K, the signal L_1 disappears and is not observed at $T < T^*$. Instead, at a lower temperature, another resonant mode L_3 was observed, which is most likely due to the presence of a small amount of paramagnetic impurity, for example, of isolated Fe^{3+} ions.

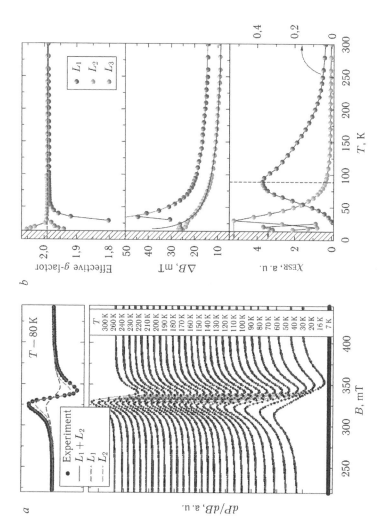

Fig. 7.36. *a*) Evolution of the ESR spectra of NaFe$_3$(HPO$_3$)$_2$(H$_2$O$_3$)$_6$ with temperature: solid lines – approximation. On the top panel is an example of the ESR spectrum decomposition into two different resonant modes. *b*) Temperature dependences of the effective *g*-factor (upper panel), linewidth (middle panel) and integrated intensity (bottom panel) for three resolved components L_1, L_2, and L_3.

The behaviour of the width of the ESR line (the middle panel in Fig. 7.36b) is similar for both allowed components L_1 and L_2. In the paramagnetic region, the width of the ESR line is practically independent of temperature and increases with decreasing temperature as it approaches the magnetic ordering temperature. Analysis in the framework of the Mori–Kawasaki–Huber critical broadening theory (2.15) leads to a satisfactory description of the experimental data for both components of L_1 and L_2 over a wide temperature range (solid lines in the middle panel of Fig. 7.36 b). The fact that appreciable broadening occurs already at a temperature of about 100 K, which is an order of magnitude higher than the ordering temperature, indicates appreciable magnetic fluctuations at high temperatures. The best agreement with the experimental data was achieved with model parameters $\Delta B_1^* = 133$ mT, $T_{N1}^{ESR} = 9.75$ K, $\beta_1 = 1.4$ and $\Delta B_2^* = 30$ mT, $T_{N2}^{ESR} = 9.89$ K, $\beta_2 = 0.47$ for the L_1 and L_2 modes, respectively. It can be seen that, despite the different temperature ranges in which $\Delta B(T)$ was approximated for the resolved components L_1 and L_2, the obtained values of T_N^{ESR} are practically identical for both resonant modes and are well comparable with the T_C obtained from the data on the magnetic susceptibility and specific heat. At the same time, the values of the critical exponentials for L_1 and L_2 differ significantly, indicating a different character of the magnetic correlations for two magnetic sublattices. It can be seen that the experimental values obtained indicate a 3D character of the magnetic fluctuations for the subsystem corresponding to the L_2 signal ($\beta_2 = 0.47$) and the low-dimensional behaviour of the subsystem corresponding to the resonance mode L_1 ($\beta_1 = 1.4$).

The data on the investigation of the Mössbauer effect are in agreement with the ESR data. In a zero magnetic field, the spectra consist of one doublet to 10 K. Below 10 K, it decomposes into two magnetic sextets and a residual doublet (Fig. 7.37), indicating the presence of two non-equivalent crystallographic positions for the iron ions. A single doublet is observed in the paramagnetic region (10–300 K), which is characterized by a quadrupole splitting $\Delta \approx 0.17$ mm/s, which corresponds to the main component of the electric field gradient (EFG), V_{zz}, of the order of 10 V/Å^2, assuming an axially symmetric EFG. This symmetry of the crystal field is apparently due to a distortion of the oxygen octahedron surrounding the iron ions in $NaFe_3(HPO_3)_2(H_2PO_3)_6$. The isomeric shift at room temperature $\delta = 0.42$ mm/s is typical of the high-spin ion Fe^{3+} ($S = 5/2$) in the octahedral oxygen coordination. The large value of

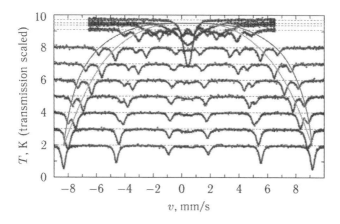

Fig. 7.37. The Mössbauer spectra in $NaFe_3(HPO_3)_2(H_2PO_3)_6$.

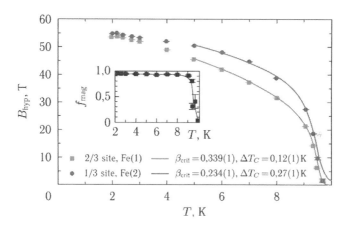

Fig. 7.38. Temperature dependences of hyperfine fields for two resolved components of Mössbauer spectra. On the inset: the fraction of magnetically ordered centres f_{mag} centres.

the hyperfine field (> 50 T) is in agreement with the presence of a high-spin Fe^{3+} ion and a negligible orbital contribution to the EFG analogously to the ESR data.

Below $T_C = 9.5$ K, the spectra break down into two distinct sextets. Moreover, the temperature dependences of the magnetic order parameter B_{hyp} show a different behaviour for the two spectral modes, as shown in Fig. 7.38. The ratio between the intensities of the partial spectra is 65:35, which agrees well with the ratio of iron ion concentrations in the crystallographic positions Fe (1) and Fe (2). A small residual paramagnetic signal (~ 5%) can be attributed to the presence of an impurity in agreement with the ESR data (L_3

Table 7.3. The exchange interactions J_i in $NaFe_3(HPO_3)_2(H_2PO_3)_6$, z_i is the number of the nearest neighbours, that is, the number of exchange paths at the unit cell site

		z_i^{Fe1}	z_i^{Fe2}	d_{Fe-Fe}, Å	J_i, K
J_1	Fe2–Fe2	0	1	4.821	0.6
J_2	Fe2–Fe2	0	1	4.943	2.4
J_3	Fe1–Fe2	2	1	5.620	2.2
J_4	Fe1–Fe2	1	1	5.788	2.0
J_5	Fe1–Fe2	2	1	5.926	0.4
J_6	Fe1–Fe2	2	1	6.317	2.3

mode in Fig. 7.36b). The approximation of $B_{hyp}(T)$ for temperatures $T \geq T_c/2$ gives the critical exponents 0.338 and 0.234 for the partial spectra of Fe (1) and Fe (2) ions, respectively (solid curves in Fig. 7.38). Thus, like the ESR data, the signal from Fe (1) shows typical 3D behaviour, while the magnetic interactions of Fe (2) ions exhibit a low-dimensional behaviour.

The isotropic exchange interactions of J_i in $NaFe_3(HPO_3)_2(H_2O_3)_6$ obtained as a result of calculations from the first principles are given in Table 7.3. Comparing the exchange interactions calculated in various supercells, it is possible to exclude any long-range exchanges that go beyond the six closest links presented in this Table. It is established that all exchanges are antiferromagnetic and include supersuper exchange through several oxygen atoms, since $Fe^{3+}O_6$ octahedra are bound only through the non-magnetic PO_3H groups. Comparing the absolute values of the exchange integrals, there is no clear dimerization within the Fe(2) sublattice or a separation between the Fe(1) and Fe(2) sublattices. Instead, there are four stronger exchanges of $\sim$ 2.0 K and two more weak exchanges of $\sim$ 0.5 K. Four strong magnetic exchanges form planes, while two weak exchanges connect these planes to the three-dimensional lattice shown in Fig. 7.39.

The interplanar connections J_2, J_3 and J_6 form triangular blocks, as shown in Fig. 7.39. Despite the frustration, they do not interfere with the establishment of the long-range magnetic order in the system, since each side of the triangle is characterized by its magnitude of interactions. One would expect that the two strongest links in the triangle will stabilize the antiparallel arrangement of the spins, while the weakest link remains with the parallel arrangement of the spins, however the case of $NaFe_3(HPO_3)_2(H_2O_3)_6$ differs from this scenario.

Here the magnetic order is stabilized by exchanges J_3 and J_6, which are slightly weaker than J_2, but they are twice as large.

The resulting spin configuration, shown in Fig. 7.39, leads to a ferrimagnetic order in the plane. The spins are parallel inside each of the sublattices Fe(1) and Fe(2), and are antiparallel between the sublattices. The interplanar links J_1 and J_5 are also frustrated by the formation of a triangle with an intraplane exchange J_4, as shown in the bottom panel of Fig. 7.39. The configuration of the ground state should be stabilized by the J_5 bond, which is weaker than J_1, but the number of such bonds is twice that, so they provide a greater stabilization energy for the magnetic configuration of the ground state. The J_5 bond is diagonal in nature, and it stabilizes the ferromagnetic order between the planes, so that the resulting moments of each layer also contribute to the formation of a common macroscopic ferrimagnetic configuration with a magnetization $M_r = 1/3 M_{sat}$, where $M_{sat} \approx 5\mu_B$/Fe is the saturation magnetization of an ion with a spin of 5/2. If, on the other hand, the interlayer order is determined by the weaker J_1, then the resulting moments of the

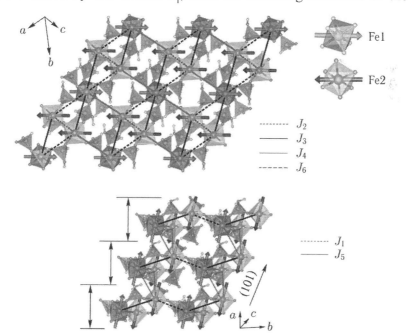

Fig. 7.39. Microscopic magnetic model of $NaFe_3(HPO_3)_2(H_2PO_3)_6$. The upper panel: magnetic planes formed by four strong bonds J_2, J_3, J_4 and J_6. The lower panel: the laying of magnetic planes in the [101] direction, the planes are connected by weaker bonds J_1 and J_5. Also shown is the ferrimagnetic ground state stabilized by bonds J_3–J_6. An arbitrary direction of the magnetic moments is selected.

neighboring layers compensate each other, and this leads to a general antiferromagnetic state. Thus, the iron phosphite lattice of iron is a ferrimagnetic layer organized by interacting diamond chains.

7.3. Shastry–Sutherland lattice

A special case of a two-dimensional frustrated lattice is a grid of orthogonal dimers $S = 1/2$, corresponding to the Shastry–Sutherland model shown schematically in the left panel of Fig. 7.40. The ground state of such a lattice was first studied in [396]. The magnetic phase diagram proposed by the authors is also shown in Fig. 7.40. The quantum ground state of such a system depends on the ratio of the intradimer interaction J_d, the frustrating interdimer interaction J, or on their ratio $\alpha = J_d/J$. When there is no intradimer interaction in the presented lattice, that is, $\alpha = 0$, then the ground state corresponds to the Néel antiferromagnetic ordering. When $\alpha \to \infty$, that is, there is only an intradimer interaction, then a spin-singlet state is realized. In the intermediate layers, because of strong frustration, the realization of non-collinear magnetic states is possible.

The realization of the two-dimensional Shastry–Sutherland lattice in the region of the spin-liquid state was observed in $SrCu_2(BO_3)_2$. The crystal structure of $SrCu_2(BO_3)_2$ has tetragonal symmetry, the space group $P4mm$ [397]. The dimers from the edge-shared CuO_4 plaquettes connected to each other by triangular groups BO_3 form the two-dimensional layers shown in the left panel of Fig. 7.41 [398].

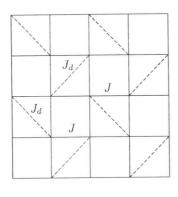
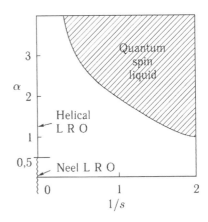

Fig. 7.40. Left panel: the Shastry–Sutherland model of a two-dimensional lattice of interacting orthogonal dimers. Right panel: magnetic phase diagram for the Shastry–Sutherland lattice [396].

These layers are separated by non-magnetic Sr^{2+} ions. The nearest copper cations Cu^{2+} $S = 1/2$, located at a distance of 2.905 Å, form magnetic dimers. The distance between the dimers is 5.132 Å, so that the exchange magnetic interaction occurs through borate groups. In each layer, not all dimers are coplanar. Vertically unfolded dimers are somewhat displaced along the c axis with respect to horizontal dimers.

In measurements of the magnetic susceptibility of $SrCu_2(BO_3)_2$, an indication of the presence of a spin gap in the spectrum of magnetic excitations was obtained [399]. The temperature dependences of the longitudinal and transverse components of the magnetic susceptibility are shown in the right panel of Fig. 7.41. In these measurements, a noticeable magnetic anisotropy of the g-factor was observed. With decreasing temperature, the susceptibility increases, reaches a maximum at 15 K, and then rapidly decreases to zero. The separation of the contributions from impurities and temperature-independent terms from the experimental $\chi(T)$ dependence made it possible to obtain the contribution of dimers. The treatment of this contribution by the function $\propto \exp(-\Delta/T)$ at low temperatures makes it possible to determine the value of the spin gap as $\Delta = 34$ K.

Approximation of the experimental data by a dimer model with a fixed maximum temperature (~ 15 K) leads to a significant discrepancy between the theoretical and experimental curves at high temperatures, as shown in the right panel of Fig. 7.41.

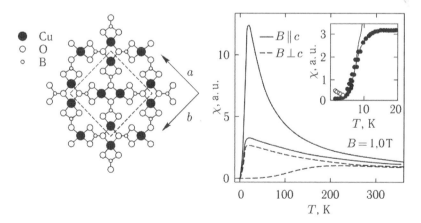

Fig. 7.41. Left panel: crystal structure of $SrCu_2(BO_3)_2$. Right panel: temperature dependences of the magnetic susceptibility of a $SrCu_2(BO_3)_2$ single crystal obtained at different directions of the magnetic field (solid and dashed line). The upper and lower solid and dashed lines show theoretical approximations in the dimer model. The inset shows the temperature dependence of the magnetic susceptibility of the matrix and its theoretical treatment [398].

Thus, the interaction between the dimers in $SrCu_2(BO_3)_2$ can not be neglected: the dimers are correlated in the layer, and each such layer is strongly frustrated. In [400] a model is proposed for the formation of a spin gap for correlated dimers, which describes well the experimental relationships. The phase diagram for the grid of orthogonal dimers obtained in the framework of this model is shown in the left panel of Fig. 7.42. At small values of J'/J, the system is described by a model of isolated dimers. With an increase in the ratio J'/J, such a two-dimensional system undergoes a first-order phase transition to a magnetically ordered state at $(J'/J)_c = 0.7$ T.

The right panel of Fig. 7.42 shows the magnetization curves of $SrCu_2(BO_3)_2$ measured in pulsed fields. Hysteresis for the input and output of a magnetic field is absent. As the magnetic field increases, the lowest triplet levels cross the level of the spin singlet at $B \sim 20$ T, which is accompanied by an increase in the magnetization. However, the field dependence of the magnetization in $SrCu_2(BO_3)_2$ demonstrates the steps, which distinguishes this metal oxide from classical spin-dimer systems, in which the magnetization increases monotonically. The plateau of the magnetization correspond to 1/8 and 1/4 of the total magnetic moment of Cu^{2+} ions. The phase boundaries for the plateau 1/8 were established as 30.1–31.7 T at $B||c$ and 26.7–28.6 T at $B \perp c$. For the plateau 1/4, the same values were 39.1–41.6 T at $B||c$ and 35.0–39.0 T at $B \perp c$. As the magnetic field increases, the system of orthogonal dimers in $SrCu_2(BO_3)_2$ successively passes through spin-gap states (plateau regions) and gapless states (between the plateaus).

The fact that triplets prefer an ordered state to a disordered one is due, apparently, to the orthogonality of the dimers closest to each

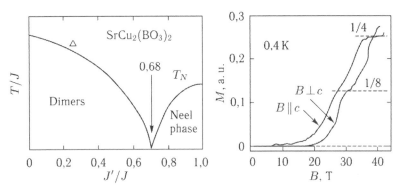

Fig. 7.42. Left panel: an empirical phase diagram for a two-dimensional grid of orthogonal dimers. The arrow marks the position of $SrCu_2(BO_3)_2$ on it. Right panel: the magnetization curves $SrCu_2(BO_3)_2$ measured in pulsed magnetic fields [398, 399].

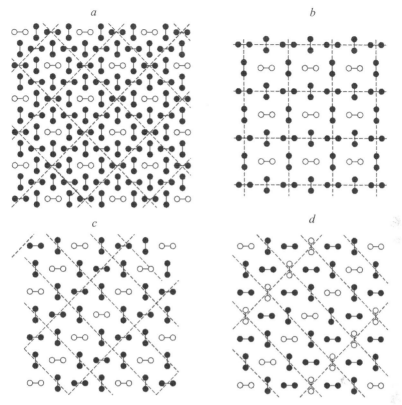

Fig. 7.43. The arrangement of singlet (closed symbols) and triplet (open symbols) states in the structure, corresponding to the destruction of singlet states in 1/8 (*a*), 1/4 (*b*, *c*) and 1/3 (*d*) of all dimers and the appearance of plateau 1/8, 1/4, 1/3 [398, 401].

other. The transition of a triplet excitation from one crystallographic position to another within the same plane is possible only in the sixth order of the perturbation theory. Taking into account the tetragonal symmetry of the $SrCu_2(BO_3)_2$ crystal, the presence of a square magnetic unit cell is a necessary condition for the formation of the ordered structure of magnetic triplets, which is the case, as shown in Fig. 7.43 for the plateau 1/8 (*a*) and for the plateau 1/4 (*b*). The same requirement is also satisfied for the plateau 1/2, 1/10, 1/16, 1/32 in a crystal with tetragonal symmetry [400, 401].

It was shown [402, 403] that triplet–triplet interactions with the next–nearest (second-order) neighbours are much weaker than interactions with more distant neighbours (the third order). This opens the possibility of forming a 1/4 plateau with stripes, as shown in Fig. 7.43 in, and not on a square lattice. Then the unit cell corresponds to a rectangular parallelepiped, in which there are no interactions with the third-order neighbours. For the 1/3 plateau, the stripe structure is shown in Fig. 7.43*d*. The presence of this plateau was confirmed in measurements of the magnetization in pulsed fields up to 69 T [404].

7.4. Potassium carbonate–manganese vanadate

The hexagonal crystal structure of the potassium carbonate–manganese vanadate $K_2Mn_3(VO_4)_2(CO_3)$, the space group $P6_3/m$, is shown in Fig. 7.44 [405]. The layered structure is a combination of modules of edge-shared MnO_6 octahedra forming a honeycomb-type lattice with VO_4 tetrahedra at the hexagon centers that alternate along the [001] direction with layers $(MnCO_3)$ formed from trigonal bipyramids MnO_5 bound through the vertices of the flat CO_3 triangles Fig. 7.45). Thus, there are two crystallographic positions in the structure for Mn^{2+} magnetic ions with different oxygen environments: in the honeycomb-type layers, the Mn1 ion is in distorted octahedra that alternate with the layers of Mn2 ions in distorted oxygen bipyramids.

The temperature dependence of the static magnetic susceptibility in the field $B = 1$ T (Fig. 7.46) in the high-temperature region

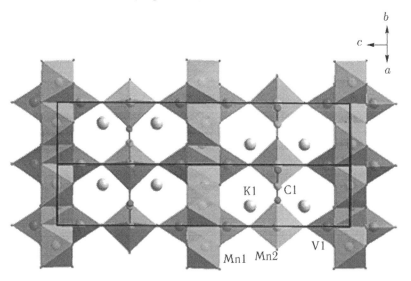

Fig. 7.44. General polyhedral view of the layered crystal structure of $K_2Mn_3(VO_4)_2(CO_3)$ in the [110] plane. The coordination polyhedra for Mn, V, and C are shown in different shades.

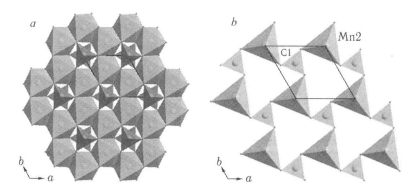

Fig. 7.45. The topology of the magnetic subsystem in the crystal structure of $K_2Mn_3(VO_4)_2(CO_3)$: *a*) layers of the first type of edge-shared MnO_6 octahedra forming honeycomb with VO_4 tetrahedra in hexagon centers; *b*) layers of the second type from the trigonal bipyramid MnO_5, connected through the vertices of the plane triangles CO_3.

satisfies the Curie–Weiss law in the presence of a temperature-independent term [197]. The best agreement with the experimental data for the approximation in the temperature range 200–300 K was obtained with the Curie constant C = 8.5 emu K/mol and the Weiss temperature Θ = −114 K. As the temperature is lowered, the experimental curve deviates significantly from the Curie–Weiss law, as shown by the line in Fig. 7.46, which suggests an increase in the role of ferromagnetic correlations in the sample under study. Moreover, at a temperature of ~ 3 K there is a sharp increase in the value of χ, possibly indicating a phase transition to the ferromagnetic state.

The left inset in Fig. 7.46 shows the temperature dependence of the magnetic susceptibility $\chi(T)$ in the field B = 0.1 T, on which a weak anomaly is observed in the region of variation of the slope $1/\chi$ at T_3 ~ 83 K. The magnetization curve $K_2Mn_3(VO_4)_2(CO_3)$ at T = 2 K as a function of the external field up to 9 T, shows a singularity of the plateau type in intermediate magnetic fields, which is replaced by a sharp increase above the critical field B_c ≈ 7 T (Fig. 7.47). This behaviour may be due to the presence of a small spontaneous magnetization in the ground state of the investigated compound.

The data on the specific heat (Fig. 7.48) are in good agreement with the data on the magnetic susceptibility and shed light on the nature of magnetic ordering. It is established that the ordering occurs in two stages with phase transitions at T_1 = 2 K and T_2 = 3 K. It is interesting to note that a comparison of the ordering

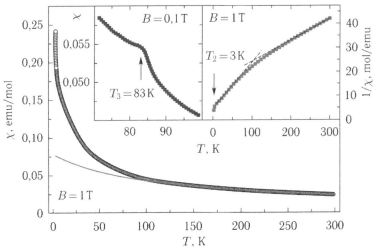

Fig. 7.46. Temperature dependence of the magnetic susceptibility of the compound $K_2Mn_3(VO_4)_2(CO_3)$ in the field $B = 1$ T. On the inserts: (on the right) the inverse magnetic susceptibility $\chi^{-1}(T)$ over a wide range of temperatures, the change in the slope at $T \sim 100$ K is shown by the dotted lines; (left) a fragment of the $\chi(T)$ in the field $B = 0.1$ T.

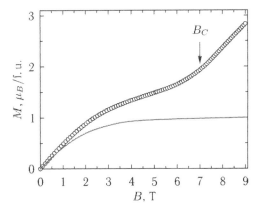

Fig. 7.47. The field dependence of the magnetization in $K_2Mn_3(VO_4)_2(CO_3)$ at $T = 2$ K. The solid line represent the Brillouin function for paramagnetic ions $S = 5/2$.

temperature with the estimated Weiss temperature shows that the frustration index ($f = \Theta/T_N$) assumes an anomalously large value in $K_2Mn_3(VO_4)_2(CO_3)$ $f \sim 50$. Similar two-step transitions were observed earlier in other low-dimensional systems and were often identified in neutron scattering experiments as the formation of a magnetic incommensurate structure at T_2, which is replaced by a commensurate phase at T_1 [226, 406, 407].

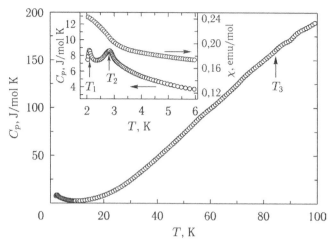

Fig. 7.48. The temperature dependence of the specific heat $C_p(T)$ in $K_2Mn_3(VO_4)_2(CO_3)$. On the inset: the low-temperature part of the dependences $C_p(T)$ and $\chi(T)$.

As can be seen from Fig. 7.48, both anomalies at $T_1 = 2$ K and $T_2 = 3$ K are observed against the background of a wide anomaly of the Schottky type, the origin of which is unclear and requires the analysis of magnetic exchange interactions. The anomaly of T_3 does not manifest itself on the temperature dependence of specific heat.

Evolution of the ESR spectra of a powder sample of $K_2Mn_3(VO_4)_2(CO_3)$ upon cooling from room temperature is shown in Fig. 7.49. ESR spectra are well described by a single line of the Lorentz-type corresponding to the exchange-narrowed line from Mn^{2+} ions. The average value of the effective g-factor at room temperature is $g = 1.978$. When cooling, the intensity of the ESR absorption line then there is a noticeable broadening and a gradual deviation from the Lorentz shape of the line, possibly indicative of the slowing down of spin-spin correlations as we approach the ordered region.

The temperature dependences of the main ESR parameters obtained from the approximation of the experimental data by the Lorentz-type function are shown on the right panel of Fig. 7.49. It can be seen from the figure that all the ESR parameters exhibit a distinct anomaly in the nature of the temperature dependences in the vicinity of $T \sim 100$ K.

Taking into account the peculiarities of the crystal structure of the test compound, it can be assumed that layers of the first type of edge-shared MnO_6 octahedra forming bee honeycombs represent the dominant magnetic subsystem, which contributes to the magnetism of $K_2Mn_3(VO_4)_2(CO_3)$. Analysis of exchange interactions in accordance

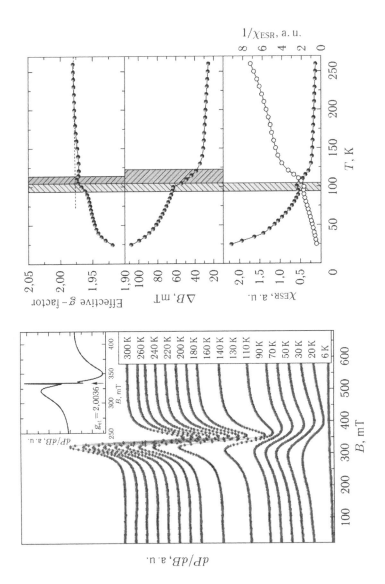

Fig. 7.49. Left panel: the evolution of the first derivative of the ESR absorption signal $K_2Mn_3(VO_4)_2(CO_3)$ with a temperature variation. On the inset: ESR spectrum, measured with a reference sample. The right panel represents the temperature dependences of the principal ESR parameters.

with the Goodenough–Kanamori rules [54] allows one to expect strong antiferromagnetic exchanges between manganese ions in layers of the first type, since each Mn1 has five half-filled orbitals on the d-shell, i.e. all t_{2g} and e_g electronic orbitals take part in the exchange, and the superexchange interaction of Mn1–O–Mn1 can be very significant. In the layers of the second type, the Mn2 ions are in a less symmetrical ligand environment of the bipyramid type, which do not have direct contacts with each other.

The Mn2 ions interact with the Mn1 ions along the supersuper-exchange path Mn1–O–V–O–Mn2, and with each other along the supersuper exhange paths Mn2–O–C–O–Mn2; in this case, both supersuper-exchanges turn out to be weaker than the super exchange Mn1–O–Mn1. This can lead to the fact that the magnetic behaviour of the subsystem Mn2 turns out to be dependent on the magnetization of the dominant subsystem Mn1. The frustration of antiferromagnetic interactions makes it difficult to form a long-range magnetic order, which is realized at temperatures very low in comparison with the energy of the exchange interactions ($J/k_B \sim \Theta$).

The nature of the anomaly detected on the temperature dependence of the magnetic susceptibility at $T_3 \sim 83$ K and in the ESR data has not been explained. Perhaps it has a local dynamic nature. Being a sensitive method of local diagnostics of the dynamic characteristics of the crystalline and spin subsystems, the ESR detects an anomaly that divides the investigated range into regions corresponding to different spin dynamics. If at high temperatures the g-factor and the linewidth remained practically temperature-independent, then in the temperature range $T < T_3$ the absorption line broadens. This is typical for the appearance of short-range order correlations, whose role increases as it approaches the ordering temperature.

The low temperatures T_1 and T_2 reflect the fact that the formation of the long-range magnetic order depends not only on the frustration of the exchange interactions in the layers of Mn1 ions with the topology of honeycombs, but also on the magnitude of the interlayer exchange interactions between Mn1 and Mn2.

7.5. Sodium–nickel phosphate hydroxide

The monoclinic crystal structure of sodium–nickel phosphate hydroxide, $Na_2Ni_3(OH)_2(PO_4)_2$, the space group $C2/m$, is shown in Fig. 7.50 [408]. The structure is a magnetically active layer of edge-

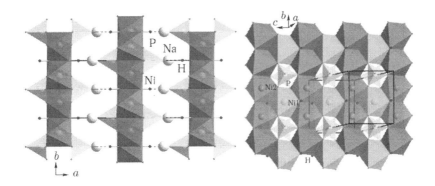

Fig. 7.50. The layered crystal structure of $Na_2Ni_3(OH)_2(PO_4)_2$ – a view in the *ab* plane (left). The organization of a magnetoactive layer with two types of edge-bound octahedra [NiO_6] and tetrahedra connecting them [PO_4] (right).

shared nickel octahedrons and phosphorus tetrahedra that alternate with non-magnetic layers of sodium polyhedra.

The layers of octahedra [$NiO_4(OH)_2$], each of which has common edges with four adjacent Ni polyhedra, are based on hexagonal close packing of oxygen ions, where the Ni^{2+} cations occupy 3/4 of the octahedral positions and can be represented as chains of octahedra formed parallel to the direction [001]. In the structure of this compound there are two types of octahedra [$NiO_4(OH)_2$], differing in symmetry type and the method of chain formation. The more symmetrical Ni1 octahedra occupy a central position within the zone, they are surrounded by four Ni2 polyhedra and do not have links with the neighbouring chains in the [010] direction. In contrast, each octahedron Ni2 has three common edges with adjacent Ni polyhedra in the same zone and one common edge with the neighbouring polyhedron Ni2 along the *b* axis (right panel of Fig. 7.50). The tetrahedrons [PO_4] have three common oxygen with nickel octahedra in each layer, whereas the fourth oxygen O4 plays the role of a hydrogen acceptor and provides hydrogen bonds between [PO_4] and adjacent Ni layers through O3–H...O4 bonds. The polyhedra [NaO_7] bound in its turn form layers parallel to the (100) plane, which alternate with the layers of [$NiO_4(OH)_2$] octahedra along the [100] direction.

The temperature dependence of the static magnetic susceptibility in the field $B = 1$ T in $Na_2Ni_3(OH)_2(PO_4)_2$ is shown in Fig. 7.51 [408].

The coincidence of the $\chi(T)$ dependences in the ZFC and FC regimes attests to the absence of spin-glass or cluster effects in

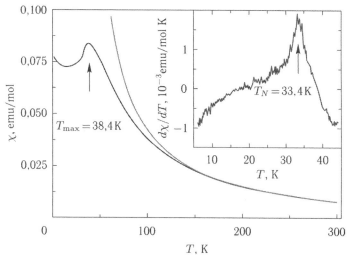

Fig. 7.51. Temperature dependence of the magnetic susceptibility of $Na_2Ni_3(OH)_2(PO_4)_2$. The line is the result of approximation by the Curie–Weiss law. On the inset, the increased low-temperature part of the derivative of the magnetic susceptibility.

the magnetization. The square of the effective magnetic moment, calculated from the value of the Curie constant $C = 3.87$ emu K/mol, is $\mu_{eff}^2 = 8C\mu_B^2 \approx 31\mu_B^2$ per formula unit. As can be seen from Fig. 7.51, the magnetic susceptibility exhibits a broad maximum at $T_{max} = 38.4$ K, while its derivative $d\chi/dT$ exhibits a sharp peak at 33.4 K, which corresponds to the Néel temperature. A wide correlation maximum in magnetic susceptibility is a sign of a reduced dimensionality of the magnetic subsystem [191]. With further cooling, the short-range order correlation regime at T_{max} is replaced by the formation of a long-range antiferromagnetic order at $T_N = 33.4$ K. The rise of the magnetic susceptibility observed at low temperatures is attributed to the presence of a small number of defects/impurities in the sample. A typical ESR spectrum of a powder sample of $Na_2Ni_3(OH)_2(PO_4)_2$ is shown in Fig. 7.52. The absorption line is relatively wide and asymmetric. Such a line shape can indicate both the anisotropy of the effective g-tensor and the possible presence of two different resonant modes. A satisfactory description of the shape of the line can be achieved by approximating the sum of two standard Lorentzians with effective g-factors $g_1 = 2.55$ (Line 1) and $g_2 = 2.18$ (Line 2), as shown in Fig. 7.52. The crystal structure of $Na_2Ni_3(OH)_2(PO_4)_2$ assumes the presence of two different the crystallographic positions of Ni1 and Ni2 for Ni^{2+} ions in the octahedral coordination (Fig. 7.50), which are in the ratio 1:2. Indeed, an estimate of the integrated

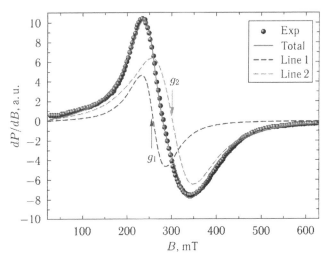

Fig. 7.52. The ESR spectrum of the powder sample $Na_2Ni_3(OH)_2(PO_4)_2$ at $T = 300$ K: the points are experimental data, the solid line is the result of approximation by the sum of two Lorentzians, the dashed lines correspond to the resolved components of the spectrum.

ESR intensity, which is proportional to the number of magnetic spins, yields a ratio of 1:2 for two resolved resonance modes. As the temperature is lowered, the effective g factor remains almost unchanged, while the width of the ESR line slightly increases for both components of the spectrum (Fig. 7.53). The total integrated intensity of the ESR is in good agreement with the behaviour of the static magnetic susceptibility, as shown in the lower part of Fig. 7.53.

Using the values of g_1 and g_2 obtained from the ESR, we can estimate the theoretical value of the effective magnetic moment

$$\mu_{theor}^2 = n_1 g_1 S(S+1)\mu_B^2 + n_2 g_2 S(S+1)\mu_B^2, \tag{7.3}$$

where $n_1 = 1$ and $n_2 = 2$, taking into account the ratio 1:2 for nickel ions in different crystallographic positions in the formula unit, and $S = 1$ for the high-spin state of Ni^{2+} (d^8). The corresponding value is $\mu_{theor}^2 \approx 32\mu_B^2 / \text{f. u.}$, which is in satisfactory agreement with the experimental estimate of $31\mu_B^2 / \text{f. u}$.

The values of superexchange interactions inside the $[NiO_4(OH)_2]$ layers and between them were estimated from the first-principles calculations described below. Qualitatively, however, we can assume that intralayer exchange interactions are responsible for the formation of the short-range order correlation regime at T_{max},

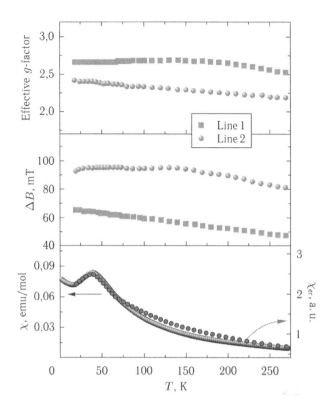

Fig. 7.53. The temperature dependences of the effective *g*-factor, the width of the ESR line for two resolved components and the integrated ESR intensity in comparison with the data for static magnetic susceptibility.

while the establishment of the long-range order at T_N is ensured by interlayer exchange interactions.

The arrangement of Ni^{2+} ions ($S = 1$) in the $[NiO_4(OH)_2]$ layers in the investigated compound $Na_2Ni_3(OH)_2(PO_4)_2$, as shown in Fig. 7.54 *b*, is unique in the sense that the topology of its magnetic subsystem is an intermediate variant between a lattice of honeycomb-type, as in the compound $K_2Mn_3(CO_3)(VO_4)_2$ (Fig. 7.54 *a*) [405], and the kagome lattice, as in compound $BaNi_3(OH)_2(VO_4)_2$ (Fig. 7.54 *c*) [409]. In the last two compounds, the magnetic subsystem is derived from a simple triangular lattice, where the voids in the center of the hexagons are connected either along the edge or along the corner, respectively. Both these lattices are strongly frustrated, so that, depending on the relationship and the sign of the exchange interactions between the nearest and next neighbours, various quantum ground states can be realized.

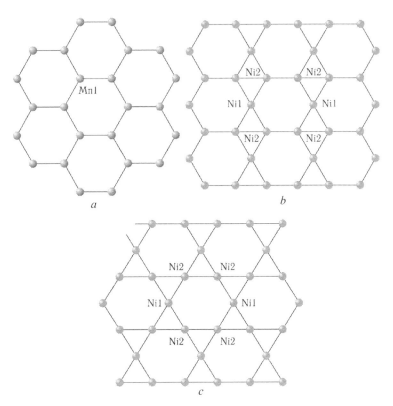

Fig. 7.54. The topologies of the magnetic subsystems $K_2Mn_3(CO_3)(VO_4)_2$ [197] with a honeycomb-type lattice (*a*), $Na_2Ni_3(OH)_2(PO_4)_2$ (*b*) and $BaNi_3(OH)_2(VO_4)_2$ with the kagome lattice [409] (*c*).

First-principles calculations showed that in the system under study there are five main exchange interactions J_1, J_2, J_3, J_4 and J_5, indicated in Fig. 7.55. The exchange integral J_3 between the edge-shared nickel octahedra at the Ni2 positions turned out to be the strongest and is about 3.3 meV (38.3 K), while the supersuper-exchanges J_4 and J_5 for Ni1–Ni1 and Ni2–Ni2 through the phosphate groups PO_4 tripled weaker in magnitude. Another dominant exchange of J_2 between the edge-shared octahedra of nickel $Ni2O_6$ turns out to be four times weaker than J_3. The exchange of J_1 between the edge-bound octahedra $Ni1O_6$ and $Ni2O_6$ is eight times weaker in magnitude than J_3. Finally, the interlayer exchange interaction between Ni1 and Ni2 located in two adjacent layers is carried out along the supersuperbial path O–H–O–P and is comparable in order of magnitude to J_1.

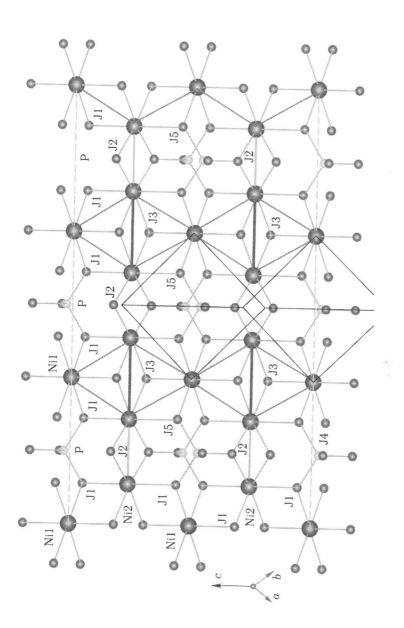

Fig. 7.55. The main paths of exchange interactions of Ni–Ni inside a magnetically active layer in $Na_2Ni_3(OH)_2(PO_4)_2$.

All the exchange integrals turned out to be antiferromagnetic, which leads to the realization of a non-trivial spin model consisting of interacting layers that are formed from the hexagons of nickel magnetic ions interacting through the strongest exchange J_3. At the same time, within the hexagon exchanges between the nearest neighbours J_1 and J_2 and diagonal exchanges through hexagon J_4 and J_5 turn out to be weaker, but also significant.

In fact, the magnetic subsystem in $Na_2Ni_3(OH)_2(PO_4)_2$ is the realization of the stripe variant of the kagome lattice with two non-equivalent positions of Ni^{2+} ions. The blocks of NiO_6 form infinite layers of hexagons that are connected along the edge along the axis b and along the corner along the c axis. In such a stripe variant in $Na_2Ni_3(OH)_2(PO_4)_2$, Ni1 positions remain unshifted for half the period relative to each other along the c axis, in contrast to the pure kagome of the $BaNi_3(OH)_2(VO_4)_2$ lattice. Accordingly, tetragonal motifs are realized between neighbouring hexagons, which are much less frustrated than triangular motifs in $BaNi_3(OH)_2(VO_4)_2$. In support of this interpretation, the fact that the ordering temperature $T_N = 33.4$ K in $Na_2Ni_3(OH)_2(PO_4)_2$ is significantly higher in comparison with other nickel phosphates [410–412] and vanadates [409].

Conclusion

Investigations in the field of low-dimensional magnetism led to the discovery of a number of new phenomena and significantly broadened and deepened the understanding of processes occurring in solids at low temperatures. The 'crystallization' of the magnetic subsystem remains, as before, the main scenario of the behaviour of magnets at low temperatures, but these states themselves are often qualitatively different from the 'classical' antiferro-, ferri- and ferromagnetic structures.

A number of inorganic compounds with a reduced dimensionality of the magnetic subsystem exhibit features of the spin-liquid, with the characteristics of this liquid being different for systems governed by the Fermi–Dirac and Bose–Einstein statistics. A key role in the development of the concept of the spin liquid was played by the Haldane model for spin chains with integer spin.

Of the most striking effects responsible for the formation of quantum ground states of matter in quasi-one-dimensional systems, one can single out the spin-Peierls transition and the formation of a spin-singlet state due to charge and orbital ordering. Common features in the behaviour of spin-liquid magnets and superconductors and the proximity to the Bose–Einstein condensation paradigm were noted at the early stages of the formation of the region of low-dimensional magnetism; however, in recent years, magnetic and nematic phenomena have come to the fore in the study of superconductivity.

Estimating the current state of research in the physics of low-dimensional magnetic systems, we can conclude that this area of science has reached the cutting edge of the development of condensed matter physics. The quantum ground states of low-dimensional magnets, established to date, are described quite well by the existing models. These models are tested on many quasi-one-dimensional and quasi-two-dimensional metaloxide compounds. Progress in this area of knowledge is connected with the search and experimental and theoretical study of new objects.

The palette of physical phenomena in the two-dimensional magnetic systems is much richer than in the one-dimensional and, especially, zero-dimensional systems. This is due to the abundance of variants of the arrangement of magnetic ions in the plane. Among these options are those in which triangular motifs support the frustration of magnetic interactions. These are, first of all, lattices such as triangle, bee honeycomb, kagome, Shastri–Sutherland and their combination. The greatest interest in the study of such objects makes it possible to realize in them a spin-liquid state, as well as exotic – chiral, helicoidal, incommensurate – magnetic structures. The theory of low-dimensional magnetism also relies heavily on the two-dimensional magnets. The Kitaev model of a honeycomb type lattice and its extensions to other triangular lattices makes it possible to obtain exact solutions for a number of spin–liquid states and allows experimental verification.

Circular and vortex structures in two-dimensional magnetic materials, called skyrmions, open up prospects for the creation of new technologies for non-volatile memory and information reading. Of exceptional interest is the topological phase transition in two-dimensional magnetic materials, the Berezinsky–Kosterlitz–Thouless transition, from the state of bound pairs of vortex-antivortex to a state with unpaired vortex formations. Conceptually, such vortices are important not only for establishing the basic mechanisms for the formation of the magnetic and superconducting states, but also for finding analogies with models used in elementary particle physics and cosmology.

References

1.   *Heisenberg W.* // Zeitschrift für Physik. 1928. V. 49. P. 619.
2.   *Dirac P. A. M.* The Principles of Quantum Mechanics. 3rd ed. Clarendon–Oxford, 1947. – 315 p.
3.   *Johnston D. C.* // In Handbook of Magnetic Materials / Ed. by K. H. J. Buschow. – Elsevier Science. Netherlands. – 1997. V. 10.
4.   *Valenti R., Saha-Dasgupta T.* // Physical Review B. 2002. V. 65. P. 144445.
5.   *Isobe M., Ueda Y.* // Journal of the Physical Society of Japan. 1996. V. 65. P. 3142.
6.   *Vasiliev A.N., Volkova O.S., Zvereva E.A., Koshelev A.V., Urusov V.S., Chareev D.A., Petkov VI, Sukhanov MV, Rahaman B., Saha-Dasgupta T.* // Physical Review B. 2015. V. 91. P. 144406 .
7.   *Clerac R., Cotton F.A., Dunbar K.R., Hillard E.A., Petrukhina M. A., Smucker B.W.,* Comptes Rendus de L Academie des Sciences Serie II Fascicule C-Chemie. 2001. V. 4. P. 315.
8.   *Choi K.-Y., Matsuda Y.H., Nojiri H.* // Physical Review Letters. 2006. V. 96. P. 107202.
9.   *Trif M., Troiani F., Stepanenko D., Loss D.* // Physical Review Letters. 2008. V. 101. P. 217201.
10. *Georgeot B., Milla F.* // Physical Review Letters. 2010. V. 104. P. 200502.
11. *Belik A., Matsuo A., Azuma M., Kindo K., Takano M.* // Journal of Solid State Chemistry. 2005. V. 178. P. 709.
12. *Bonner J. C., Fisher M.E.,* Physical Review. 1964. V. 135. P. A640.
13. *Mo X., Etheredge K. M. S., Hwu S.- J., Huang Q.* // Inorganic Chemistry. 2006. V. 45. P. 3478.
14. *Yasui Y., Kawamura Y., Kobayashi Y., Sato M.* // Journal of Applied Physics. 2014. V. 115. P. 17E125.
15. *Drechsler S. L., Richter J., Gippius A. A., Vasiliev A., Bush A. A., Moskvin A. S., Ma'lek J., Prots Yu., Schnelle W., Rosner H.* // Europhysics Letters. 2006. V. 73. P. 83.
16. *Taniguchi S., Nishikawa T., Yasui Y., Kobayashi Y., Sato M., Nishioka T.* // Journal of the Physical Society of Japan. 1995. V. 64. P. 2758.
17. *Ulutagay-Kartin M., Hwu S.-J., Clayhold J. A.* // Inorganic Chemistry. 2003. V. 42. P. 2405.
18. *Belik A., Azuma M., Matsuo A., Whangbo M.-H., Koo H.-J., Kikuchi J., Kaji T., Okubo S., Ohta H., Kindo K., Takano M.,* // Inorganic Chemistry. 2005. V. 44. P. 6632.
19. *Weihong Z., Oitmaa J., Hamer C. J.* // Physical Review B. 1998. V. 58. P. 14147.
20. *Demokritov S. O., Demidov V. E., Dzyapko O., Melkov G. A., Serga A. A., Hillebrands, B., Slavin A. N.* // Nature. 2006. V. 443. P. 430.
21. *Bloch F.* // Zeitschrift fuer Physik. 1930. V. 61. P. 206.

22. *Matsubara T., Matsuda H.* // Prog. Theor. Phys. 1956. V. 16. P. 569.

23. *Giamarchi T., Ruegg C., Tchernyshyov O.* // Nature Physics. 2008. V. 4. P. 198.

24. *Sasago Y., Uchinokura K., Zheludev A., Shirane G.* // Physical Review B. 1997. V. 55. P. 8357.

25. *Sebastian S. E., Harrison N., Batista C. D., Balicas L., Jaime M., Sharma P. A., Kawashima N., Fisher I. R.* // Nature. 2006. V. 441. P. 617.

26. *Liu, G., Greedan G. E.* // Journal of Solid State Chemistry. 1994. V. 108. P. 267.

27. *Vasiliev A., Volkova O., Zvereva E., Isobe M., Ueda Y., Yoshii S., Nojiri H., Mazurenko V., Valentyuk M., Anisimov V., Solovyev I., Klingeler R., Büchner B.* // Physical Review B. 2013. V. 87. P. 134412.

28. *Andersen O. K.* // Physical Review B. 1975. V. 12. P. 3060.

29. *Andersen O. K., Jepsen O.* // Physical Review Letters. 1984. V. 53. P. 2571.

30. *Korotin M. A., Elfimov I. S., Anisimov V. I., Troyer M., Khomskii D. I.* // Physical Review Letters. 1999. V. 83. P. 1387.

31. *Bethe H. A.* // Zeitschrift für Physik. 1931. V. 71. P. 205.

32. *Fisher M.E.* // Physica. 1960. V. 26. P. 618.

33. *Johnston D. C., Kremer R. K., Troyer M., Wang X., Klümper A., Bud'ko S. L., Panchula A. F., Canfield P. C.* // Physical Review B. 2000. V. 61. P. 9558.

34. *Ising E.* // Zeitschrift für Physik. 1925. V. 31. P. 253.

35. *Ishii R., Gautreaux D., Onuma K., Machida Y., Maeno Y., Nakatsuji S., Chan J. Y.* // J Journal of the American Chemical Society. 2010. V. 132. P. 7055.

36. *Weeks C., Song Ya., Suzuki M., Chernova N. A., Zavalij P. Y.* // Journal of Materials Chemistry. 2003. V. 13. P. 1420.

37. *Koo H. J., Whangbo M. H.* // Solid State Sciences. 2010. V. 12. P. 685.

38. *Feyerherm R., Abens S., Günther D., Ishida T., Meißner M., Meschke M., Nogami T., Steiner M.* // Journal of Physics: Condensed Matter. 2000. V. 12. P. 8495.

39. *Köppen M., Lang M., Helfrich R., Steglich F., Thalmeier P., Schmidt B., Wand B., Pankert D., Benner H., Aoki H., Ochiai A.* // Physical Review Letters. 1999. V. 82. P. 4548.

40. *Eggert S., Affleck I. Takahashi M* // Physical Review Letters. 1994. V. 73. P. 332.

41. *Abragam A., Blini B.* Electron paramagnetic resonance of transition ions / A. Abragam, B. Blini; edited by S.A. Altshuler and G.V. Skrotskii. – Moscow: Mir. – 1972. 2 T.

42. *Carrington A., McLachlan A.D.* Introduction to Magnetic Resonance. Chapman & Hall. – London, 1967 – 266 p.

43. Bertaina S., Pashchenko V. A., Stepanov A., Masuda T., *Uchinokura K.* // Physical Review Letters. 2004. V. 92. P. 057203.

44. *Oshikawa M., Affleck I.* // Physical Review Letters. 1997. V. 79. P. 2883.

45. *Affleck I., Oshikawa M.* // Physical Review B. 1999. V. 60. P. 1038.

46. *Oshikawa M., Affleck I.* // Physical Review Letters. 1999. V. 82. P. 5136.

47. *Oshikawa M., Affleck I.* // Physical Review B. 2002. V. 65. P. 134410.

48. *Kwek L. C., Takahashi Y., Choo K. W.* // Journal of Physics: Conference Series. 2009. V. 143. P. 012014.

49. *Sun Z., Wang X. G., Li Y. Q.* // New Journal of Physics. 2005. V. 7. P. 83.

50. *Liu P., Liang M.-L., Yuan B.* // European Physical Journal D-Atomic, Molecular, Optical and Plasma Physics. 2007. V. 41. P. 571.

51. *Gisin N.* // Physics Letters A. 1991. V. 154. P. 201.

52. *Majumdar C. K., Ghosh D. K.* // Journal of Mathematical Physics. 1969. V. 10. P. 1388.

53. *Drechsler S.-L., Volkova O., Vasiliev AN, Tristan N., Richter J., Schmitt M., Rosner H., Malek J., Klingeler R., Zvyagin AA, Büchner B.* // Physical Review Letters. 2007. V. 98. P. 077202.

54. *Goodenough, J. B.* Magnetism and Chemical Bond, Robert E. Krieger Publishing Company, Huntington. – N.Y., 1976. – 939 p.

55. *Klingeler R.* private communication. 2016.

56. *Tarui Y., Kobayashi Y., Sato M.* // Journal of the Physical Society of Japan. 2008. V. 77. P. 043703.

57. *Yasui Y., Igawa N., Kakurai K.* // JPS Conference Proceedings. 2015. V. 8. P. 034012.

58. *Hibble S. J., Köhler J., Simon A., Paider S.* // Journal of Solid State Chemistry. 1990. V. 88. P. 534.

59. *Tams G., Mullerbuchbaum H.* // Journal of Alloys and Compounds. 1992. V. 189. P. 241.

60. *Berger R., Önnerud P., Tellgren R.* // Journal of Alloys and Compounds. 1992. V. 184. P. 315.

61. *Berger R., Meetsma A., van Smaalen S.* // Journal of the Less Common Metals. 1991. V. 175. P. 119.

62. *Paszkowicz W., Marczak M., Vorotynov A. M., Sablina K. A., Petrakovski G. A.* // Powder diffraction. 2001. V. 16. P. 30.

63. *Bush AA, Kamentsev K. Ye., Tishchenko E. A.* // Inorganic materials. 2003. V. 39. P. 1.

64. *Maljuk A., Kulakov A. B., Sofin M., Capogna L., Strempfer J., Lin C. T., Jansen M., Keimer B.* // Journal of Crystal Growth. 2004. V. 263. P. 338.

65. *Markina M., Chistyakova T., Tristan N., Buchner B., Bush A., Vasiliev A.* // JETP. 2007. P. 132. P. 27.

66. *Masuda T., Zheludev A., Bush A., Markina M., Vasiliev A.* // Physical Review Letters. 2004. V. 92. P. 177201.

67. *Masuda T., Zheludev A., Roessli B., Bush A., Markina M., Vasiliev A.* // Physical Review B. 2005. V. 72. P. 014405.

68. *Gippius A. A., Morozova E. N., Moskvin A. S., Zalessky A. V., Bush A. A., Baenitz M., Rosner H., Drechsler S.-L.* // Physical Review B. 2004. V. 70. P. 020406.

69. *Gippius A. A., Morozova E. N., Moskvin A. S., Drechsler S.-L., Baenitz M.* // Journal of Magnetism and Magnetic Materials. 2006. V. 300. P. e335.

70. *Capogna L., Mayr M., Horsch P., Raichle M., Kremer RK, Sofin M., Maljuk A., Jansen M., Keimer B.* // Physical Review B. 2005. V. 71. P. 140402R.

71. *Rusydi A., Mahns I., Mueller S., Ruebhausen M., Park S., Choi YJ, Zhang CL, Cheong S.-W., Smadici S., Abbamonte P., van Zimmermann M., Sawatzky GA* // Applied Physics Letters. 2008. V. 92. P. 262506.

72. *Park S., Choi Y. J., Zhang C.L., Cheong S.-W.* // Physical Review Letters. 2006. V. 98. P. 057601.

73. *Seki S., Yamasaki Y., Soda M., Matsuura M., Hirota K., Tokura Y.* // Physical Review Letters. 2008. V. 100. P. 127201.

74. *Hsu H.C., Lin J.-Y., Lee W.L., Chu M.-W., Imai T., Kao Y.J., Hu C.D., Liu H.L., Chou F.C.* // Physical Review B. 2010. V. 82. P 094450.

75. *Zhao L., Yeh K.-W., Rao SM, Huang T.-W., Wu P., Chao W.- H., Ke C.-T., Wu C.-E., Wu M .-K.* // Europhysics Letters. 2012. V. 97. P. 37004.

76. *Mostovoy M.* // Physical Review Letters. 2006. V. 96. P. 067601.

77. *Kimura T., Goto T., Shintani H., Ishizaka K., Arima T., Tokura Y.* // Nature (London). 2003. V. 426. P. 55.

78. *Hur N., Park S., Sharma P. A., Ahn J. S., Guha S., Cheong S.-W.* // Nature (London). 2004. V. 429. P. 392.

79. *Yakubovich O. V., Kiriukhina G. V., Dimitrova O. V., Zvereva E. A., Shvanskaya L. V., Volkova O. S., Vasiliev A.N.* // Dalton Transaction. 2016. V. 45. P. 2598.

80. *Isobe M., Ninomiya E., Vasil'ev A.N., Ueda Y.* // Journal of the Physical Society of Japan. 2002. V. 71. P. 1423.

81. *Vasil'ev A.N., Voloshok T. N., Ignatchik O. L., Isobe M., Ueda Y.* // JETP Letters. 2002. V. 76.P. 35.

82. *Shvanskaya L., Yakubovich O., Bychkov A., Shcherbakov V., Golovanov A., Zvereva E., Volkova O., Vasiliev A.* // Journal of Solid State Chemistry. 2015. V. 222. P. 44.

83. *Wertz J.E, Bolton J.R.,* Electron Spin Resonance: Elementary Theory and Practical Applications. McGraw-Hill Book Company, New York, 1972 – 500 p..

84. *Choukroun J., Pashchenko A., Ksari Y., Henry JY, Mila F., Millet P., Monod P., Stepanov A., Dumas J., Buder R.* // European Physical Journal B-Condensed Matter and Complex Systems. 2000. V. 14. P. 655.

85. *Pashchenko V. A., Sulpice A., Mila F., Millet P., Stepanov A., Wyder P.* // European Physical Journal B-Condensed Matter and Complex Systems. 2001. V. 21. P. 473.

86. *Chabre F., Ghorayeb A. M., Millet P., Pashchenko V., Stepanov A.* // Physical Review B. 2005. V. 72. P. 012415.

87. *Deisenhofer J., Schaile S., Teyssier J., Wang Z., Hemmida M., Krug von Nidda H.-A., Eremina RM, Eremin MV, Viennois R., Giannini E., Marel D., Loidl A.* // Physical Review B. 2012. V. 86. P. 214417.

88. *Ivanshin V. A., Yushankhai V., Sichelschmidt J., Zakharov D. V., Kaul E. E., Geibel C.* // Physical Review B. 2003. V. 68. P. 044404.

89. *Misra S. K. Sun J.-S* // Physical Review B. 1990. V. 42. P. 8601.

90. *Jain V. K., Vugman N. V. Yadav V.S.* // Physical Review B. 1988. V. 37. P. 9716.

91. *Krishna R. M., Gupta S. K.* // Bulletin of Magnetic Resonance. 1994. V. 16. P. 239.

92. *Kataev V., Choi K.-Y., Grüninger M., Ammerahl U., Buchner B., Freimuth A. Revcolevschi* // Physical Review Letters. 2001. V. 86. P. 2882.

93. *Wolter AUB, Lipps F., Schäpers M., Drechsler S.-L., Nishimoto S., Vogel R., Kataev V., Büchner B., Rosner H., Schmitt M., Uhlarz M ., Skourski Y., Wosnitza J., Süllow S., Rule KC* // Physical Review B. 2012. V. 85. P. 014407.

94. *Gatteschi D., Bencini A.* Electron Paramagnetic Resonance of Exchange Coupled Systems / D. Gatteschi, A. Bencini. – Heidelberg: Springer. – 1990. – 286 p.

95. *Rakitin Yu.V., Kalinnikov V.T.,* Modern magnetochemistry / Ed. V. I. Nefedova. – St. Petersburg, Nauka. – 1994. – 276 p.

96. *Uhrig G. S., Normand B.* // Physical Review B. 2001. V. 63. P. 134418.

97. *Tsirlin A. A., Nath R., Abakumov A. M., Shpanchenko R. V., Geibel C., Rosner H.* // Physical Review B. 2010. V. 81. P. 174424.

98. *Doyle R. P., Bauer T., Julve M., Lloret F., Cano J., Nieuwenhuyzen M., Kruger P. E.* // Dalton Transactions. 2007. V. 44. P. 5140.

99. *Venegas-Yazigi, D., Mun~oz-Becerra K., Brown K., Aliaga C., Kniep R., Cardoso-Gil R., Schnelle W., Paredes-García V., Aguirre P., Spodine. E.* // Polyhedron. V. V. 29. P. 2426.

100. *Berdonosov P. S., Kuznetsova E. S., Dolgikh V. A., Sobolev A. V., Presniakov I. A., Olenev A. V., Rahaman B., Saha-Dasgupta T., Zakharov K. V., Zvereva E. A., Volkova O. S., Vasiliev A. N.* // Inorganic Chemistry. 2014. V. 53. P. 5830.

101. *Fisher M.E.* // American Journal of Physics. 1964. V. 32. P. 343.

102. *Dormann E., Jaccarino V.* // Physics Letter A. 1974. V. 48. P. 81.

103. *Kawasaki K.* // Progress of Theoretical Physics. 1968. V. 39. P. 285.

104. *Kawasaki K.* // Physics Letter A. 1968. V. 26. P. 543.

105. *Mori H., Kawasaki K.* // Progress of Theoretical Physics. 1962. V. 28. P. 971.

106. *Huber D. L.* // Physical Review B. 1972. V. 6. P. 3180.

107. *Drechsler S.-L., Richter J., Kuzian R., Malek J., Tristan N., Buechner B., Moskvin*

AS, *Gippius AA, Vasiliev A., Volkova O., Prokofiev A., Rakoto H. , Broto J.-M., Schnelle W., Scmitt M., Ormeci A., Loison C., Rosner H.* // Journal of Magnetism and Magnetic Materials. 2007. V. 316. P. 306.

108. *Hatfield W. E.* // Journal of Applied Physics. 1980. V. 52. P. 1985.

109. *Kikuchi J., Motoya K., Yamauchi T., Ueda Y.* // Physical Review B. 1999. V. 60. P. 6731.

110. *Garrett A. W., Nagler S. E., Tennant D. A., Sales B. C., Barnes T.* // Physical Review Letters. 1997. V. 79. P. 745.

111. *Yamauchi T., Narumi Y., Kikuchi J., Ueda Y., Tatani K., Kobayashi T. C., Kindo K., Motoya K.* // Physical Review Letters. 1999. V. 83. P. 3729.

112. *Haldane F. D. M.* // Physical Review Letters. 1983. V. 50. P. 1153.

113. *Haldane F. D. M.* // Physical Review Letters. 1988. V. 61. P. 1029.

114. *Botet R., Jullien R., Kolb M.* // Physical Review B. 1983. V. 28. P. 3914.

115. *Botet R., Jullien R.* // Physical Review B. 1983. V. 27. P. 613.

116. *Law J. M., Benner H., Kremer R. K.* // Journal of Physics: Condensed Matter. 2013. V. 25. P. 065601.

117. *Yamamoto S., Miyashita S.* // Physical Review B. 1993. V. 48. P. 9528.

118. *Uchiyama Y., Sasago Y., Tsukada I., Uchinokura K., Zheludev A., Hayashi T., Miura N., Böni P.* // Physical Review Letters. 1999. V. 83. P. 632.

119. *Zheludev A., Masuda T., Uchinokura K., Nagler S. E.* // Physical Review B. 2001. V. 64. P. 134415.

120. *Sakai T., Takahashi M.* // Physical Review B. 1990. V. 42. P. 4537.

121. *Bera A. K., Lake B., Islam A. T. M. N., Klemke B., Faulhaber E., Law J. M.* // Physical Review B. 2013. V. 87. P. 224423.

122. *Tsujii N., Suzuki O., Suzuki H., Kitazawa H., Kido G.* // Physical Review B. 2005. V. 72. P. 104402.

123. *Pahari B., Ghoshray K., Sarkar R., Bandyopadhyay B., Ghoshray A.* // Physical Review B. 2006. V. 73. P. 012407.

124. *Maeda Y., Hotta C., Oshikawa M.* // Physical Review Letters. 2007. V. 99. P. 057205.

125. *Vollenkle H., Wittmann A., Nowotny H.* // Monatshefte für Chemie. 1967. V. 98. P. 1352.

126. *Hase M., Terasaki I., Uchinokura K.* // Physical Review Letters. 1993. V. 70. P. 3651.

127. *Nishi M., Fujita O., Akimitsu J.* // Physical Review B. 1994. V. 50. P. 6508.

128. *Lorenz T., Ammerahl U., Ziemes R., Buchner B., Revcolevschi A., Dhallen G.* // Physical Review B. 1996. V. 55. P. 15610.

129. *Winkelmann H., Gamper E., Buchner B., Braden M., Revcolevschi A., Dhalenne G.* // Physical Review B. 1995. V. 51. P. 12884.

130. *Takehana K., Oshikiri M., Kido G., Hase M., Uchinokura K.* // Journal of the Physical Society of Japan. 1996. V. 65. P. 2783.

131. *Hirota K., Cox D. E., Lorenzo J. E., Shirane G., Tranquada J. M., Hase M., Uchinokura K., Kojima H., Shibuya Y., Tanaka I.* // Physical Review Letters. 1994. V. 73. P. 736.

132. *Hase M., Terasaki I., Uchinokura K., Tokunaga M., Miura N., Obara H.* // Physical Review B. 1993. V. 48. P. 9616.

133. *Ohta H., Imagawa S., Ushiroyama H., Motokawa M., Fujita O., Akimitsu J.* // Journal of the Physical Society of Japan. 1994. V. 63. P. 2870.

134. *Nojiri H., Shimamoto Y., Miura N., Hase M., Uchinokura K., Kojima H., Tanaka I., Shibuya Y.* // Physical Review B. 1995. V. 52. P. 12749.

135. *Kiryukhin V., Keimer B., Hill J. P., Vigliante A.* // Physical Review Letters. 1996. V. 76. P. 4608.

136. *Lorenz T., Büchner B., van Loosdrecht P. H. M., Schoennfeld F., Chouteau G., Revcolevschi A., Dhalenne G.* // Physical Review Letters. 1998. V. 81. P. 148.

137. *Ohashi H., Fujita T., Osawa T.* // Journal of the Japanese Association of Mineralogists, Petrologists and Economic Geologists. 1982. V. 77. P. 305.

138. *White S. R., Noack R. M., Scalapino D. J.* // Physical Review Letters. 1994. V. 73. P. 886.

139. *Cabra D. C., Honecker A., Pujol P.* // Physical Review Letters. 1997. V. 79. P. 5126.

140. *Cabra D. C., Honecker A., Pujol P.* // Physical Review B. 1998. V. 58. P. 6241.

141. *Azzouz M., Shahin K., Chitov G. Y.* // Physical Review B. 2007. V. 76. P. 132410.

142. *Azuma M., Hiroi Z., Takano M., Ishida K., Kitaoka Y.* // Physical Review Letters. 1994. V. 73. P. 3463.

143. *Thurber K. R., Imai T., Saitoh T., Azuma M., Takano M., Chou F. C.* // Physical Review Letters. 2000. V. 84. P. 558.

144. *Barnes T., Dagotto E., Riera J., Swanson E. S.* // Physical Review B. 1993. V. 47. P. 3196.

145. *Gopalan S., Rice T. M., Sigrist M.* // Physical Review B. 1994. V. 49. P. 8901.

146. *Noack R. M., White S., Scalapino D.* // Physical Review Letters. 1994. V. 73. P. 882.

147. *Azzouz M., Chen L., Moukouri S.* // Physical Review B. 1994. V. 50. P. 6233.

148. *Troyer M., Tsunetsugu H., Wurtz D.* // Physical Review B. 1994. V. 50. P. 13515.

149. *Larochelle S., Greven M.* // Physical Review B. 2004. V. 69. P. 092408.

150. *Totsuka K., Suzuki M.* // Journal of Physics: Condensed Matter. 1995. V. 7. P. 6079.

151. *Hiroi Z., Azuma M., Takano M., Bando Y.* // Journal of Solid State Chemistry. 1991. V. 95. P. 230.

152. *Ishida K., Kitaoka Y., Asayama K., Azuma M., Hiroi Z., Takano M.* // Journal of the Physical Society of Japan. 1994. V. 63. P. 3222.

153. *Azuma M., Takano M., Eccleston R. S.* // Journal of the Physical Society of Japan. 1998. V. 67. P. 740.

154. *Johnston D.C, Troyer M., Miyahara S., Lidsky D., Ueda K., Azuma M., Hiroi Z., Takano M., Isobe M., Ueda Y., Korotin MA, Anisimov VI, Mahajan AV, Miller LL* Magnetic Susceptibilities of Spin-1/2 Antiferromagnetic Heisenberg Ladders and Applications to Ladder Oxide Compounds // arXiv: cond-mat / 0001147.

155. *Magishi K., Matsumoto S., Kitaoka Y., Ishida K., Asayama K., Ueda M., Nagata T., Akimitsu J.* // Physical Review B. 1998. V. 57. P. 11533.

156. *Ishida K., Kitaoka Y., Tokunaga Y., Matsumoto S., Azuma M., Hiroi Z., Takano M.* // Physical Review B. 1996. V. 53. P. 2827.

157. *Isobe M., Ueda Y.* // Journal of the Physical Society of Japan. 1996. V. 65. P. 1178.

158. *Ohama T., Yasuoka H., Isobe M., Ueda Y.* // Physical Review B. 1999. V. 59. P. 3299.

159. *von Schnering H.-G., Grin Y. U., Kaupp M., Samer M., Kremer R. K., Jepsen O., Chatterji T., Weiden M.* // Zeitschrift für Kristallographie-Crystalline Materials. 1998. V. 213. P. 246.

160. *Cuoco M., Horsch P., Mack F.* // Physical Review B. 1999. V. 60. P. R8438.

161. *Vasil'ev A.N., Markina M.M., Kagan M. Yu., Isobe M., Ueda Yu.* // Pis'ma Zh. Teor. Fiz. 2001. V. 73. P. 401.

162. *Ravy S., Jegoudez J., Revcolevschi A.* // Physical Review B. 1999. V. 59. P. R681.

163. *Mostovoy M. V., Khomskii D. I.* // Solid State Communications. 2000. V. 113. P. 159.

164. *Fujii Y., Nakao H., Yoshihama T., Nishi M., Nakajima K., Kakurai K., Isobe M., Sawa H., Ueda Y.* // Journal of the Physical Society of Japan. 1997. V. 66. P. 326.

165. *Sawa H., Ninomiya E., Ohama T., Nakao H., Ohwada K., Murakami Y., Fujii Y., Noda Y., Isobe M., Ueda Y.* // Journal of the Physical Society of Japan . 2002. V. 71.

P. 385.

166. *Ohwada K., Fujii Y., Katsuki Y., Muraoka J., Nakao H., Murakami Y., Sawa H., Ninomiya E., Isobe M., Ueda Y.* // Physical Review Letters. 2005. V. 94. P. 106401.

167. *McCarron III E. M., Subramanian M. A., Calabrese J. C., Harlow R. L.* // Material Research Bulletin. 1988. V. 23. P. 1355.

168. *Matsuda M., Katsumata K.* // Physical Review B. 1996. V. 53. P. 12201.

169. *Takigawa M., Motoyama N., Eisaki H., Uchida S.* // Physical Review B. 1998. V. 57. P. 1124.

170. *Eccleston R. S., Uehara M., Akimitsu J., Eisaki H., Motoyama N., Uchida S.* // Physical Review Letters. 1998. V. 81. P. 1702.

171. *Cox D. E., Iglesias T., Hirota K., Shirane G., Matsuda M., Motoyama N., Eisaki H., Uchida S.* // Physical Review B. 1998. V. 57. 10750.

172. *Ueda Y., Isobe M., Yamauchi T.* // Journal of Physics and Chemistry of Solids. 2002. V. 63. P. 951.

173. *Yamada H., Ueda Y.* // Journal of the Physical Society of Japan. 1999. V. 68. P. 2735.

174. *Yamauchi T., Ueda Y., Mori N.* // Physical Review Letters. 2002. V. 89. P. 057002.

175. *Onoda H., Takahashi T., Nagasava H.* // Journal of the Physical Society of Japan. 1982. V. 51. P. 3868.

176. *Presura C., Popincius M., van Loosdrecht P. H. M., van der Marel D., Mostovoy M., Yamauchi T. Ueda Y* // Physical Review Letters. 2003. V. 90. P. 026402.

177. *Vasil'ev A.N., Marchenko V.I., Smirnov A.I., Sosin S.S., Yamada H., Ueda Y.* // Physical Review B. 2001. V. 64. P. 174403.

178. *Markina M., Klimov K., Vasiliev A. N., Freimut A., Kordonis K., Krainer M., Lorentz T., Yamauchi T., Ueda Yu.* // Letters in the Zh. 2004. 76. P. 670.

179. *Suzuki T., Yamauchi I., Shimizu Y., Itoh M., Takeshita N., Terakura C., Takagi H., Tokura Y., Yamauchi T., Ueda Y.* // Physical Review B. 2009. V 79. P. 081101R.

180. *Bednorz, J.G., Mueller, K.A.*, Zeitschrift für Physik B 1986. V. 64. P. 1893.

181. *Mermin N.D., Wagner H.* // Physical Review Letters. 1966. V. 17. P. 1133.

182. *Neves E. J., Perez J.F.* // Physics Letters A. 1986. V. 114. P. 331.

183. *Manousakis E.* // Review of Modern Physics. 1991. V. 63. P. 1.

184. *White S. R., Chernyshev A.L.* // Physical Review Letters. 2007. V. 99. P. 127004.

185. *Mourigal M., Zhitomirsky M.E., Chernyshev A.L.* // Physical Review B. 2010. V. 82. P. 144402.

186. *Seabra L., Sindzingre P., Momoi T., Shannon N.* // Physical Review B. 2016. V. 93. P. 085132.

187. *Kosterlitz, J.M., Thouless, D.J., Journal of Physics C: Solid State Physics. 1973. V. 6. P. 1181.*

188. *Kosterlitz J.M.* // Journal of Physics C: Solid State Physics. 1974. V. 7. P. 1046.

189. *Berezinskii V. L.* // Soviet Physics Journal of Experimental and Theoretical Physics. 1971. V. 32. P. 493.

190. *Berezinskii V. L.* // Soviet Physics Journal of Experimental and Theoretical Physics. 1971. V. 34. P. 610.

191. *de Jongh L. J., Miedema A.R.* // Advances in Physics. 1974. V. 23. P. 1.

192. *Cuccoli A., Roscilde T., Vaia R., Verrucchi P.,* / Physical Review Letters. 2003. V. 90. P. 167205.

193. *Cuccoli A., Roscilde T., Tognetti V., Vaia R., Verrucchi P.,* // Physical Review B. 2003. V. 67. P. 104414.

194. *Cuccoli A., Roscilde T., Vaia R., Verrucchi P.* // Physical Review B. 2003. V. 68. P. 060402.

195. *Regnault L.P., Rossat-Mignod J., Henry J.Y., de Jongh L.J.* // Journal of Magnetism and Magnetic Materials. 1983. V. 31–34. P. 1205.

196. *Gaveau P., Boucher J.P., Regnault L.P., Henry Y.* // Journal of Applied Physics. 1991. V. 69. P. 6228.

197. *Förster T., Garcia F.A., Gruner T., Kaul E.E., Schmidt B., Geibel C., Sichelschmidt J.* // Physical Review B. 2013. V. 87. P. 180401.

198. *Shpanchenko R.V., Kaul E.E., Geibel C., Antipov E.V.* // Acta Crystallographica Section C: Crystal Structure Communications. 2006. V. 62. P. i88.

199. *Nalbandyan V.B., Zvereva E.A., Yalovega G.E., Shukaev I.L., Ryzhakova A.P., Guda A.A., Stroppa A., Picozzi S., Vasiliev A.N., Whangbo M.-H.* // Inorganic Chemistry. 2013. V. 52. P. 11850.

200. *Law J. M., Reuvekamp P., Glaum R., Lee C., Kang J., Whangbo M.-H., Kremer R. K.* // Physical Review B. 2011. V. 84. P. 014426.

201. *Muller O., White W.B., Roy R.* // Zeitschrift für Kristallographie-Crystalline Materials. 1969. V. 130. 1. P. 112.

202. *Pernet M., Quezel G., Coing-Boyat J., Bertaut E.-F.* // Bulletin de la Société française de minéralogie et de cristallographie. 1969. V. 92. P. 264.

203. *Seferiadis N., Oswald H.R.* // Acta Crystallographica Section C: Crystal Structure Communications. 1987. V. 43. P. 10.

204. *Whangbo M.-H., Koo H. J., Dai D.* // Journal of Solid State Chemistry. 2003. V. 176. P. 417.

205. *Dai D., Whangbo M.-H.* // The Journal of Chemical Physics. 2001. V. 114. P. 2887.

206. *Sachdev S. The quantum phase of matter* // arXiv: 1203.4565v4.

207. *Yafet Y., Kittel C.* // Physical Review. 1952. V. 87. P. 290.

208. *Kawamura H., Miyashita S.* // Journal of the Physical Society of Japan. 1985. V. 54. P. 4530.

209. *Chubukov A. V., Golosov D. I.* // Journal of Physics: Condensed Matter. 1991. V. 3. P. 69.

210. *Seabra L., Momoi T., Sindzingre P., Shennon N.* // Physical Review B. 2011. V 84. P. 214418.

211. *Farnell, D.J., J., Zinke, R., Schulenburg, J., Richter, J.,* Journal of Physics: Condensed Matter. 2009. V. 21. P. 406002.

212. *Wannier G. H.* // Physical Review. 1950. V. 79. P. 357.

213. *Anderson, P. W.,* Materials Research Bulletin. 1973. V. 8. P. 153.

214. *Chen R., Ju H., Jiang H.-C., Starykh O. A., Balents L.* // Physical Review B. 2013. V. 87. P. 165123.

215. *Zvereva EA, Nalbandyan VB, Evstigneeva MA, Koo H.-J., Whangbo M.-H., Ushakov AV, Medvedev BS, Medvedeva LI, Gridina NA, Yalovega GE, Churikov AV, Vasiliev AN, Büchner B* // Journal of Solid State Chemistry. 2015. V. 225. P. 89.

216. *Richards P. M.* // Solid State Communications. 1973. V. 13. P. 253.

217. *Anders A. G., Volotski S. V.* // Journal of Magnetism and Magnetic Materials. 1983. V. 31. P. 1169.

218. *Shirata Y., Tanaka H., Matsuo A., Kindo K.* // Physical Review Letters. 2012. V. 108. P. 057205.

219. *Susuki T., Kurita N., Tanaka T., Nojiri H., Matsuo A., Kindo K., Tanaka H.* // Physical Review Letters. 2013. V. 110. P. 267201.

220. *Melchy P. E., Zhitomirsky M. E.* // Physical Review B. 2009. V. 80. P. 064411.

221. *Doi Y., Hinatsu Y., Ohoyama K.* // Journal of Physics: Condensed Matter. 16. V. 8923.

222. *Shannon R. D., Rogers D. B., Prewitt C.T.,* Inorganic Chemistry. 1971. V. 10. P. 713.

223. *Seki S., Onose Y., Tokura Y.* // Physical Review Letters 2008 V. 101. P. 067204.

224. *Kimura K., Nakamura H., Kimura S., Hagiwara M., Kimura T.* // Physical Review Letters. 2009 V. 103. P. 107201.

225. *Kimura K., Nakamura H., Ohgushi K., Kimura T.* // Physical Review B. 2008 V. 78. P. 140401.

226. *Vasiliev A., Volkova O., Presniakov I., Baranov A., Demazeau G., Broto JM, Millot M., Leps N., Klingeler R., Buchner B., Stone MB, Zheludev A.* / / Journal of Physics: Condensed Matter. 2010 V. 22. P. 016007.

227. *Terada N., Khalyavin D. D., Manuel P., Tsujimoto Y., Knight K., Radaelli P. G., Suzuki H. S., Kitazawa H.* // Physical Review Letters 2012. V. 109. P. 097203.

228. *Kimura T., Lashley J. C., Ramirez A. P.* // Physical Review B. 2006 V. 73. P. 220401.

229. *Nakajima T., Mitsuda S., Takahashi K., Yamano M., Masuda K., Yamazaki H., Prokes K., Kiefer K., Gerischer S., Terada N., Kitazawa H., Matsuda M., Kakurai K., Kimura H., Noda Y., Soda M., Matsuura M., Hirota K.* // Physical Review B. 2009 V. 79. P. 214423.

230. *Kanetsuki S., Mitsuda S., Nakajima T., Anazawa D., Katori H. A., Prokes K.* // Journal of Physics: Condensed Matter 2007 V. 19. P. P. 145244.

231. *Seki S., Yamasaki Y., Shiomi Y., Iguchi S., Onose Y. Tokura Y* // Physical Review. 2007. V. 75. P. 100403.

232. *Terada N., Nakajima T., Mitsuda S., Kitazawa H., Kaneko K., Metoki N.* // Physical Review B. 2008. V. 78. P. 014101.

233. *Singh K., Maignan A., Martin C., Simon C.* // Chemistry of Materials. 2009. V. 21. P. 5007.

234. *Streltsov S. V., Poteryaev A. I., Rubtsov A. N.* // Journal of Physics: Condensed Matter 2015. V. 27. P. 165601.

235. *Poienar M., Vecchini C., Andrée G., Daoud-Aladine A., Margiolaki I., Maignan A., Lappas A., Chapon L., Hervieu M., Damay F., Martin C.* / / Chemistry of Material. 2011. V. 23. P. 85.

236. *Ushakov A. V., Streltsov S. V., Khomskii D. I.* // Physical Review B. 2014. V. 89. P. 024406.

237. *Okuda T., Kishimoto T., Uto K., Hokazono T., Onose Y., Tokura Y., Kajimoto R., Matsuda M.* // Journal of the Physical Society of Japan. 2009. V. 7. P. 013604.

238. *Kawamura H., Miyashita S.* // Journal of the Physical Society of Japan. 1984. V. 53. P. 4138.

239. *Okubo T., Kawamura H.* // Journal of the Physical Society of Japan. 2010. V. 79. P. 084706.

240. *Halperin B. I., Nelson D. R.* // Physical Review Letters. 1978. V. 41. P. 121.

241. *Hemmida M., Krug von Nidda H.-A., Loidl A.* // Journal of the Physical Society of Japan. 2011 V. 80. P. 053707.

242. *Hemmida M., Krug von Nidda H.-A., Büttgen N., Loidl A., Alexander LK, Nath R., Mahajan AV, Berger RF, Cava RJ, Singh Y., Johnston DC* // Physical Review B. 2009. V. 80. P. 054406.

243. *Olariu A., Mendels P., Bert F., Alexander L. K., Mahajan A. V., Hillier A. D., Amato A.* // Physical Review B. 2009. V. 79. P. 224401.

244. *Rossler, U.K., Bogdanov, A. N., Pfleiderer, C.,* Nature. 2006. V. 442. P. 797.

245. *Yu X. Z., Onose Y., Kanazawa N., Park J.H., Han J. H., Matsui Y., Nagaosa N., Tokura Y.* // Nature. 2010. V. 465. P. 901.

246. *Pfleiderer C., Rosch A.* // Nature. 2010. V. 465. P. 880.

247. *Mühlbauer S., Binz B., Jonietz F., Pfleiderer C., Rosch A., Neubauer A., Georgii R., Böni. P.* // Science. 2009. V. 323. P. 915.

248. *Adams T., Chacon A., Wagner M., Bauer A., Brandl G., Pedersen B., Berger H., Lemmens P., Pfleiderer C.* // Physical Review Letters. 2012. V. 108. P. 237204.

249. *Marty K., Simonet V., Ressouche E., Ballou R., Lejay P., Bordet P.* // Physical Review Letters. 2008. V. 101. P. 247201.

250. *Stock C., Chapon L.C., Schneidewind A., Su Y., Radaelli P.G., McMorrow D.F., Bombardi A., Lee N., Cheong S.-W.* / / Physical Review B. 2011. V. 83. P. 104426.

251. *Johnson R.D., Cao K., Chapon L.C., Fabrizi F., Perks N., Manuel P., Yang J.J., Oh Y. S., Cheong S.-W., Radaelli. P.G.* // Physical Review Letters. 2013. V. 111. P. 017202.

252. *Reimers, J.N., Greedan, J.E., Subramanian M.A.*, Journal of Solid State Chemistry. 1989. V. 79. P. 263.

253. *Werner J., Koo C., Klingeler R., Vasiliev A.N., Ovchenkov Yu.A, Polovkova A.S., Raganyan G.V., Zvereva E.A.* // Physical Review B. 2016. V. 94. P. 104408.

254. *Nalbandyan V.B., Zvereva EA, Nikulin A. Yu., Shukaev IL, Whangbo M.-H., Koo H.-J., Abdel-Hafiez M., Chen X.-J., Koo C., Vasiliev AN , Klingeler R.* // Inorganic Chemistry. 2015. V. 54. P. 1705.

255. *Fisher M. E.* // Proceedings of the Royal Society. 1960. V. 66. P. 254.

256. *Fisher M.E.* // Philosophical Magazine. 1962. V. 7. P. 1731.

257. *Koo H. J., Whangbo M.-H.* // Inorganic Chemistry. 2014. V. 53. 7. P. 3812.

258. *Zvereva E. A., Savelieva O. A., Titov Ya. D., Evstigneeva MA, Nalbandyan VB, Kao CN, Lin J.-Y., Presniakov IA, Sobolev AV, Ibragimov SA, Abdel-Hafiez M., Krupskaya Yu., Jähne C., Tan G.*, P. 1550.

259. *Leidl R., Klingeler R., Büchner B., Holtschneider M., Selke W.* // Physical Review B. 2006. V. 73. P. 224415.

260. *Mulder A., Ganesh R., Capriotti L., Paramekant A.* // Physical Review B. 2010. V. 81. P. 214419.

261. *Li P.Y. Y., Bishop R. F., Farnell D.J.J., Campbell C.E.* // Physical Review B. 2012. V. 86. P. 144404.

262. *Oitmaa J., Singh R.R.P.* // Physical Review B. 2011 V. 84. P. 094424.

263. *Clark B.K., Abanin D.A., Sondhi S.L.* // Physical Review Letters. 2011. V. 107. P. 087204.

264. *Lee Y.-W., Yang M.-F.* // Physical Review B. 2012. V. 85. P. 100402.

265. *Kitaev A.* // Annals of Physics. 2006. V. 321. P. 2.

266. *Kimchi I., Analytis J. G., Vishwanath A.* // Physical Review B. 2014. V. 90. P. 205126.

267. *Chaloupka J., Jackeli G., Khaliullin G.* // Physical Review Letters. 2010. V. 105. P. 027204.

268. *Jiang H.-C., Gu Zh.-Ch .., Qi X.-L .., Trebst S.* // Physical Review B. 2011. V. 83. P. 245104.

269. *Singh Y., Gegenwart P.* // Physical Review B. 2010. V. 82. P. 064412.

270. *Singh Y., Manni S., Reuther J., Berlijn T., Thomale R., Ku W., Trebst S., Gegenwart P.* // Physical Review Letters. 2012. V. 108. P. 127203.

271. *Ye F., Chi S., Cao H., Chakoumakos BC, Fernandez-Baca JA, Custelcean R., Qi TF, Korneta OB, Cao G.* // Physical Review B. 2012. V. 85. P. 180403 .

272. *Biffin A., Johnson R. D., Kimchi I., Morris R., Bombardi A., Analytis J. G., Vishwanath A., Coldea R.* // Physical Review Letters. 2014. V. 113. P. 197201.

273. *Plumb K. W., Clancy J. P., Sandilands L. J., Shankar V. V., Hu Y. F., Burch K. S., Kee H.-Y., Kim Y.-J.* / / Physical Review B. 2014. V. 90. P. 041112.

274. *Johnson R.D., Williams S.C., Haghighirad A.A., Singleton J., Zapf V., Manuel P., Mazin I.I., Li Y., Jeschke H.O., Valentí R., Coldea R.* // Physical Review B. 2015. V. 92. P. 235119.

275. *Sears J.A., Songvilay M., Plumb K.W., Clancy J. P., Qiu Y., Zhao Y., Parshall D., Kim Y.-J.* // Physical Review B. 2015 V. 91. P. 144420.

276. *Streltsov S., Mazin I. I., Foyevtsova K.* // Physical Review B. 2015. V. 92. P. 134408.

277. *Singh, D.J.*, Physical Review, B. 2015. V. 91. P. 214420.

278. *Kimber S.A.J., Mazin I.I., Shen J., Jeschke H.O., Streltsov S.V., Argyriou D.N., Valenti R., Khomskii D.I.* // Physical Review B. 2014. V. 89. P. 081408.

279. *Smirnova O., Azuma M., Kumada N., Kusano Y., Matsuda M., Shimakawa Y., Takei T., Yonesaki Y., Kinomura N.* // Journal of the American Chemical Society. 2009. V. 131. P. 8313.

280. *Matsuda M., Azuma M., Tokunaga M., Shimakawa Y., Kumada N.* // Physical Review Letters. 2010. V. 105. P. 187201.

281. *Heinrich M., Krug von Nidda H.-A., Loidl A., Rogado N., Cava R.J.* // Physical Review Letters. 2003. V. 91. P. 137601.

282. *Rogado N., Huang Q., Lynn J. W., Ramirez A. P., Huse D., Cava R.J.* // Physical Review B. 2002. V. 65. P. 144443.

283. *Nagarajan R., Uma S., Jayaraj M. K., Tate J., Sleight A. W.* // Solid State Sciences. 2002. V. 4. P. 787.

284. *Smirnova O. A., Nalbandyan V. B., Petrenko A. A., Avdeev M.* // Journal of Solid State Chemistry. 2005. V. 178. P. 1165.

285. *Politaev V. V., Nalbandyan V. B., Petrenko A. A., Shukaev I. L., Volotchaev V. A., Medvedev B. S.* // Journal of Solid State Chemistry. 2010. V. 183. P. 684.

286. *Evstigneeva M. A., Nalbandyan V. B., Petrenko A. A., Medvedev B. S., Kataev A. A.* // Chemistry of Materials. 2011. V. 23. P. 1174.

287. *Kumar V., Gupta A., Uma S.* // Dalton Transaction. 2013. V. 42. P. 14992.

288. *Nalbandyan V. B., Avdeev M., Evstigneeva M. A.* // Journal of Solid State Chemistry. 2013. V. 199. P. 62.

289. *Skakle, J. M. S., Castellanos, M. A. R., Tovar S. T., West A. R.* // Journal of Solid State Chemistry. 1997. V. 131. P. 115.

290. *Nalbandyan V. B., Petrenko A. A., Evstigneeva M. A.* // Solid State Ionics. 2013. V. 233. P. 7.

291. *Ma X., Kang K., Ceder G., Meng Y. S.* // Journal of Power Sources. 2007. V. 173. P. 550.

292. *Roudebush, J. H., Cava R. J.,* // Journal of Solid State Chemistry. 2013. V. 204. P. 178.

293. *Gupta A., Mullins C. B., Goodenough J. B.* // Journal of Power Sources. 2013. V. 243. P. 817.

294. *Ma J., Bo S.-H., Wu L., Zhu Y., Gray C. P., Khalifah P. G.* // Chemistry of Materials. 2015. V. 27. P. 2387.

295. *Viciu L., Huang Q., Morosan E., Zandbergen H. W., Greenbaum N. I., McQueen T., Cava R. J.* // Journal of Solid State Chemistry. 2007. V. 180. P. 1060.

296. *Schmidt W., Berthelot R., Sleight A. W., Subramanian M. A.* // Journal of Solid State Chemistry. 2013. V. 201. P. 178.

297. *Berthelot R., Schmidt W., Sleight A. W., Subramanian M. A.* // Journal of Solid State Chemistry. 2012. V. 196. P. 225.

298. *Kumar V., Bhardwaj N., Tomar N., Thakral V., Uma S.* // Inorganic Chemistry. 2012. V. 51. P. 10471.

299. *Berthelot R., Schmidt W., Muir S., Eilertsen J., Etienne L., Sleight A. W., Subramanian M. A.* // Inorganic Chemistry. 2012. V. 51. P. 5377.

300. *Roudebush, J.H., Andersen N.H., Ramlau R., Garlea V. O., Toft-Petersen R., Norby P., Schneider R., Hay J. N., Cava R. J.* // Inorganic Chemistry. 2013. V. 52. P. 6083.

301. Seibel, E. M., Roudebush, J. H., Wu, H., Huang Q., Ali, M. N., Ji H., Cava R. J., Inorganic Chemistry. 2013. V. 52. P. 13605.

302. *Roudebush, J. H., Sahasrabudhe G., Bergman S. L., Cava R. J.* // Inorganic Chemistry. 2015. V. 54. P. 3203.

303. *Miura Y., Hirai R., Kobayashi Y., Sato M.* // Journal of the Physical Society of Japan. 2006. V. 75. P. 084707.

304. *Miura Y., Yasui Y., Moyoshi T., Sato M., Kakurai K.* // Journal of the Physical Society of Japan. 2008. V. 77. P. 104709.

305. *Miura Y., Hirai R., Fujita T., Kobayashi Y., Sato M.* // Journal of Magnetism and Magnetic Materials. 2007. V. 310. P. 389.

306. *Kuo C. N., Jian T. S., Lue C. S.* // Journal of Alloys and Compounds. 2012. V. 531. P. 1.

307. *Derakhshan S., Cuthbert H. L., Greedan J. E., Rahaman B., Saha-Dasgupta T.* // Physical Review B. 2007. V. 76. P. 104403.

308. *Xu J., Assoud A., Soheinia N., Derakhshan S., Cuthbert H. L., Greedan J. E., Whang-bo M. H., Kleinke H.* // Inorganic Chemistry. 2005. V. 44. P. 5042.

309. *Morimoto K., Itoh Y., Yoshimura K., Kato M., Hirota K.* // Journal of the Physical Society of Japan. 2006. V. 75. P. 083709.

310. *Sankar R., Panneer Muthuselvam I., Shu G. J., Chen W. T., Karna Sunil K., Jayavel R., Chou F. C.* // CrystEngComm. 2014. V. 16. P. 10791.

311. *Schmitt M., Janson O., Golbs S., Schmidt M., Schnelle W., Richter J., Rosner H.* // Physical Review B. 2014. V. 89. P. 174403.

312. *Climent-Pascual E., Norby P., Andersen N. H., Stephens P. W., Zandbergen H. W., Larsen J., Cava R. J.*, Inorganic Chemistry. 2012. V. 51. P. 557.

313. *Bhardwaj N., Gupta A., Uma S.* // Dalton Transactions. 2014. V. 43. P. 12050.

314. *Schmidt W., Berthelot R., Etienne L., Wattiaux A., Subramanian M. A.* // Material Research Bulletin. 2014. V. 50. P. 292.

315. *Zvereva E. A., Evstigneeva M. A., Nalbandyan V. B., Savelieva O. A., Ibragimov S. A., Volkova O. S., Medvedeva L. I., Vasiliev A. N., Klingeler R., Büchner B.* // Dalton Transaction. 2012. V. 41. P. 572.

316. *Zvereva EA, Stratan MI, Ovchenkov YA, Nalbandyan VB, Lin J.-Y., Vavilova EL, Iakovleva MF, Abdel-Hafiez M., Silhanek AV, Chen X.-J., Stroppa A., Picozzi S., Jeschke HO, Valenti R., Vasiliev AN* // Physical Review B. 2015. V. 92. P. 144401.

317. *Willett R. D., Waldner F.* // Journal of Applied Physics. 1982. V. 53. P. 2680.

318. *Willett R. D., Wong R. J.* // Journal of Magnetic Resonance. 1981. V. 42. P. 446.

319. *Castner T. G. Seehra Jr, Seehra M.S.* // Physical Review B. 1971. V. 4. P. 38.

320. *Castner T. G., Seehra M. S.* // Physical Review B. 1993. V. 47. P. 578.

321. *Zorko A., Bert F., Ozarowski A., van Tol J., Boldrin D., Wills A. S., Mendels P.* // Physical Review B. 2013. V. 88. P. 144419.

322. *Huber D. L., Seehra M. S.* // Journal of Physics and Chemistry of Solids. 1975. V. 36. P. 723.

323. *Seehra M. S., Ibrahim M. M., Babu V. S., Srinivasan G.* // Journal of Physics: Condensed Matter. 1996. V. 8. P. 11283.

324. *Zorko A., Arèon D., van Tol H., Claude Brunel L., Kageyama H.* // Physical Review B. 2004. V. 69. 17. P. 174420.

325. *Watson MD, McCollam A., Blake SF, Vignolles D., Drigo L., Mazin II, Guterding D., Jeschke HO, Valenti R., Ni N., Cava R., Coldea AI* // Physical Review B. 2014. V. 89. P. 205136.

326. *Zvereva E. A., Stratan M. I., Ushakov A. V., Nalbandyan V. B., Shukaev I. L., Silhanek A. V., Abdel-Hafiez M., Streltsov S. V., Vasiliev A. N.* // Dalton Transaction. 2016. V. 45. P. 7373.

327. *Streltsov S. V., Khomskii D. I.* // Physical Review B. 2008. V. 77. P. 064405.

328. *Streltsov S. V., Popova O. A., Khomskii D. I.* // Physical Review Letters. 2006. V. 96. P. 249701.

329. *Okamoto K., Tonegawa T., Takahashi Y., Kaburagi M.* // Journal of Physics: Condensed Matter. 2003. V. 15. P. 5979.

330. *Ishii M., Tanaka H., Mori M., Uekusa H., Ohashi Y., Tatani K., Narumi Y. Kindo K //* Journal of the Physical Society of Japan. 2000 V. 69. P. 340.

331. *Okamoto K., Tonegawa T., Takahashi Y. Kaburagi M,* Journal of Physics: Condensed Matter 1999. V. 11. P. 10485.

332. *Takano K., Kubo K. Sakamoto H //* Journal of Physics: Condensed Matter 1996. V. 8. P. P. 6405.

333. *Yan S., Huse D. A., White S. R. //* Science. 2011. V. 332. P. 1173.

334. *Sindzingre P., Misguich G., Lhuillier C., Bernu B., Pierre L., Waldtmann Ch., Everts H.-U. //* Physical Review Letters. 2000. V. 84. P. 2953.

335. *Shores M. P., Nytko E. A., Bartlett B. M., Nocera D. G. //* Journal of the American Chemical Society. 2005. V. 127. P. 13462.

336. *Imai T., Nytko E. A., Bartlett B. M., Shores M. P., Nocera D. G. //* Physical Review Letters. 2008. V. 100. P. 077203.

337. *Helton JS, Matan K., Shores MP, Nytko EA, Bartlett BM, Yoshida Y., Takano Y., Suslov A., Qiu Y., Chung J.-H., Nocera DG, Lee YS //* Physical Review Letters. 2007. V. 98. P. 107204.

338. *Okamoto Y., Yoshida H., Hiroi Z. //* Journal of the Physical Society of Japan. 2009. V. 78. P. 033701.

339. *Lafontaine M. A., Bail A. L., F'erey G. //* Journal of Solid State Chemistry. 1990. V. 85. P. 220.

340. *Colman R.H., Bert F., Boldrin D., Hillier A.D., Manuel P., Mendels P., Wills A.S. //* Physical Review B. 2011. V. 83. P. 180416.

341. *Yoshida H., Michiue Y., Takayama-Muromachi E., Isobe M. //* Journal of Material Chemistry. 2012. V. 22. P. 18793.

342. *Quilliam J. A., Bert F., Colman R. H., Boldrin D., Wills A. S., Mendels P. //* Physical Review B. 2011. V 84. P. 180401.

343. *Yoshida M., Okamoto Y., Yoshida H., Takigawa M., Hiroi Z. //* Journal of the Physical Society of Japan. 2013. V. 82. P. 013702.

344. *Valldor M., Andersson M. //* Solid State Sciences. 2002. V. 4. P. 923.

345. *Valldor M. //* Journal of Physics: Condensed Matter. 2004. V. 16. P. 9209.

346. *Soda M., Yasui Y., Moyoshi T., Sato M., Igawa N., Kakurai K. //* Journal of the Physical Society of Japan. 2006. V. 75. P. 054707.

347. *Chapon L. C., Radaelli P. G., Zheng H., Mitchell J. F. //* Physical Review B. 2006. V. 74. P. 172401.

348. *Valldor M. //* Solid State Sciences. 2006. V. 8. P. P. 1272.

349. *Schwenka W., Valldor M., Lemmens P. //* Physical Review Letters. 2007. V. 98. P. 067201.

350. *Nakayama N., Mizota T., Ueda Y., Sokolov A. N., Vasiliev A. N. //* Journal of Magnetism and Magnetic Materials. 2006. V. 300. P. 98.

351. *Sheptyakov D. V., Podlesnyak A., Barilo S. N., Shiryaev S. V., Bychkov G. L., Khalyavin D. D., Chernyshov D. Yu., Leonyuk N. I. //* PSI Scientific Report. 2001. V. 3. P. 64.

352. *Bychkov G. L., Shiryaev S. V., Soldatov A. G., Barilo S. N., Sheptyakov D. V., Conder K., Pomjakushina E., Podlesnyak A., Furrer A., Bruetsch R. //* Crystal Research and Technology. 2005. V. 40. P. 395.

353. *Gatal'skaya VI, Shiryaev SV, Bychkov GL, Dabkowska H., Dube P., Greedan J. E. //* Collected papers of the Institute of Solid State Physics, Belarus. 2005. P. 204.

354. *Markina M., Vasiliev A. N., Nakayama N., Mizota T., Ueda Y. //* Journal of Magnetism and Magnetic Materials. 2010. V. 322. P. 1249.

355. *Caignaert V., Pralong V., Maignan A., Raveau B. //* Solid State Communications. 2009. V. 149. P. 453.

356. *Singh K., Caignaert V., Chapon L. C., Pralong V., Raveau B., Maignan A.* // Physical Review B. 2012. V. 86. P. 024410.

357. *Caignaert V., Maignan A., Singh K., Simon Ch., Pralong V., Raveau B., Mitchell JF, Zheng H., Huq A., Chapon LC* // Physical Review B. 2013. V. 88 P. 174403.

358. *Lee N., Vecchini C., Choi Y. J., Chapon L. C., Bombardi A., Radaelli P. G., Cheong S.-W.* // Physical Review Letters. 2013. V. 110. P. 137203.

359. *Mill' BV, Butashin AV, Khodzhabagyan GG, Belokoneva Ye. L., Belov NV* / / Dokl. AN SSSR. 1982. V. 246. P. 1385.

360. *Mill B. V., Pisarevsky Yu. V.* // Proceedings of the IEEE / EIA International Frequency Control Symposium, Kansas City, USA. 2000. P. 133.

361. *Puccio, D., Saldanho, N., Malocha, D. C. Pereira da Cunha, M* // Proceedings of the IEEE International Frequency Control Symposium, New Orleans, USA. 2002. P. 324.

362. *Yokota Y., Yoshikawa A., Futami Y., Sato M., Tota K., Onodera K., Yanagida T.* // IEEE Transactions on Ultrasonics, Ferroelectrics, and Frequency Control. 2012. V. 59. P. 1868.

363. *Belokoneva E.L., Belov N.V.* // Dokl. AN SSSR. 1981. P. 260. P. 1363.

364. *Marty K., Bordet P., Simonet V., Loire M., Ballou R., Darie C., Kljun J., Bonville P., Isnard O., Lejay P., Zawilski B., Simon C.* / / Physical Review B. 2010. V. 81. P. 054416.

365. *Zhou H. D., Wiebe C. R., Jo J. J., Balicas L., Urbano R. R., Lumata L. L., Brooks J. S., Kuhns P. L., Reyes A. P., Qiu Y., Copley J. R. D., Gardner J. S.* // Physical Review Letters. 2009. V. 102. P. 067203.

366. *Bordet P., Gelard I., Marty K., Ibanez A., Robert J., Simonet V., Canals B., Ballou R., Lejay P.* // Journal of Physics: Condensed Matter. 2006. V. 18. P. 5147.

367. *Robert J., Simonet V., Canals B., Ballou R., Bordet P., Lejay P., Stunault A.* // Physical Review Letters. 2006. V. 96. P. 197205.

368. *Zhou H. D., Vogt B. W., Janik J. A., Jo Y.-J., Balicas L., Qiu Y., Copley J. R. D., Gardner J. S., Wiebe C. R.* // Physical Review Letters. 2007. V. 99. P. 236401.

369. *Zorko A., Bert F., Mendels P., Bordet P., Lejay P., Robert J.* // Physical Review Letters. 2008. V. 100. P. 147201.

370. *Zhou H. D., Wiebe C. R., Yo Y. J., Balicas L., Takano Y., Case M. J., Qiu Y., Copley J. R. D., Gardner J. S.* Nanoscale freezing of the 2D spin liquid $Pr_3Ga_5SiO_{14}$ // arXiv: 0808.2819v1.

371. *Lumata LL, Besara T., Kuhns PL, Reyes AP, Zhou HD, Wiebe CR, Balicas L., Jo YJ, Brooks JS, Takano Y., Case MJ, Qiu Y., Copley JRD, Gardner JS, Choi KY , Dalal NS, Hoch MJR* // Physical Review B. 2010. V. 81. P. 224416.

372. *Mill' B. B.* // Zh. Inorg. Khimii. 2002. V. 47. P. 812.

373. *Sharma A. Z., Silverstein H. J., Hallas A. M., Luke G. M., Wiebe C. R.* // Journal of Solid State Chemistry. 2016. V. 233. P. 14.

374. *Marty K., Simonet V., Bordet P., Ballou R., Lejay P., Isnard O., Ressouche E., Bourdarot F., Bonville P.* // Journal of Magnetism and Magnetic Materials. 2009. V. 321. P. 1778.

375. *Zhou H. D., Lumata L. L., Kuhns P. L., Reyes A. P., Choi E. S., Dalal N. S., Lu J., Jo Y. J., Balicas L., Brooks J. S., Wiebe C. R.* // Chemistry of Materials. 2009. V. 21. P. 156.

376. *Lee C., Kan E., Xiang H., Whangbo M.-H.* // Chemistry of Materials. 22. V. P. 5291.

377. *Chaix L., de Brion S., Le'vy-Bertrand F., Simonet V., Ballou R., Canals B., Lejay P.* // Physical Review Letters. 2013. V. 110. P. 157208.

378. *Mill' B. B.* // Zh. Inorg. Khimii. 2009. V. 54. P. 1270.

379. *Ivanov V. Y., Mukhin A. A., Prokhorov A. S., Mill' B. V.* // Solid State Phenomena.

2009. V. 152-153. P. 299.

380. *Silverstein, H. J., Cruz-Kan K., Hallas A. M., Zhou H. D., Donaberger R. L., Hernden B. C., Bieringer M., Choi E. S., Hwang J. M., Wills A. S.* // Chemistry of Materials. 2012. V. 24. P. 664.

381. *Krizan J. W., de la Cruz C., Andersen N. H. Cava R.J* // Journal of Solid State Chemistry. 2013. V. 203. P. 310.

382. *Silverstein H. J., Sharma A. Z., Stoller A.J., Cruz-Kan K., Flacau R., Donaberger R. L., Zhou H. D., Manuel P., Huq A., Kolesnikov A. I., Wiebe C. R.* // Journal of Physics: Condensed Matter. 2013. V. 25. P. 246004.

383. *Markina M. M., Mill B. V., Zvereva E. A., Ushakov A. V., Streltsov S. V., Vasiliev A. N.* // Physical Review B. 2014. V. 89. P. 104409.

384. *Markina M. M., Mill B. V., Ovchenkov YA, Zvereva E. A., Vasiliev A. N.* // Physics and Chemistry of Minerals. 2016. V. 43. P. 51.

385. *Silverstein H. J., Sharma A. Z., Cruz-Kan K., Zhou H. D., Huq A., Flacau R., Wiebe C. R.* // Journal of Solid State Chemistry. 2013. V. 204. P. 102.

386. *Silverstein H. J., Huq A., Lee M. S., Choi E. S., Zhou H. D., Wiebe C. R.* // Journal of Solid State Chemistry. 2015. V. 221. P. 216.

387. *Pchelkina Z. V., Streltsov S. V.* // Physical Review B. 2013. V. 88. P. 054424.

388. *Lowe, W.* Paramagnetic Resonance in Solids, Ed. G. V. Skrotsky. – Moscow: Foreign Literature. – 1962. – 242 p.

389. *Hall T. P. P., Hayes W.* // Journal of Chemical Physics. 1960. V. 32. P. 1871.

390. *Van Stapele R. P., Henning J. C. M., Hardeman G. E. G., Bongers P. F.* // Physical Review. 1966. V. 150. P. 310.

391. *Van Stapele R. P., Beljers H. G., Bongers P. F., Zijlstra H.* // The Journal of Chemical Physics. 1966. V. 44. P. 3719.

392. *Willett R. D., Gatteschi D., Kahn O.* (Eds) Magneto-structural correlations in exchange coupled systems // NATO AS1 Series. C. 140. – Reidel D. .: Dodrecht. The Netherlands. – 1985.

393. *Munaò I., Zvereva E. A., Volkova O. S., Vasiliev A. N., Armstrong R. A., Lightfoot P.* // Inorganic Chemistry. 2016. V. 55. P. 2558.

394. *Vasiliev AN, Volkova OS, Zvereva EA, Ovchenkov EA, Munao I., Armstrong R., Lightfoot P., Vavilova EL, Kamusella S., Klauss H.-H., Werner J., Koo C., Klingeler R ., Tsirlin AA* / Physical Review B. 2016. V. 93. P. 134401.

395. *Yamada I.* // Journal of the Physical Society of Japan. 1972. V. 33. P. 979.

396. *Shastry B. S., Sutherland B.* // Physica B. 1981. V. 108. P. 1069.

397. *Smith R. W., Keszler D. A.* // Journal of Solid State Chemistry. 1991. V. 93. P. 430.

398. *Kageyama H., Vasiliev A. N.* // Nature. 2002. № 2. P. 21.

399. *Kageyama H., Yoshimura K., Stern R., Mushnikov N. V., Onizuka K., Kato M., Kosuge K., Slichter C. P., Goto T., Ueda Y.* // Physical Review Letters. 1999. V. 82. P. 3168.

400. *Miyahara S., Ueda K.* // Physical Review Letters. 1999. V. 82. P. 3701.

401. *Miyahara S., Ueda K.* // Physical Review B. 2000. V. 61. P. 3417.

402. *Chen S., Han B.* // European Physical Journal B-Condensed Matter and Complex Systems. 2003. V. 31. P. 63.

403. *Miyahara S., Ueda K.* // Journal of Physics: Condensed Matter V. 15. R327 2003.

404. *Kageyama H., Ueda Y., Narumi Y., Kindo K., Kosaka M., Uwatoko Y.* // Progress of Theoretical Physics Supplements. 2002. V. 145. P. 17.

405. *Yakubovich O. V., Yakovleva E. V., Golovanov A. N., Volkov A. S., Volkova O. S., Zvereva E. A., Dimitrova O. V., Vasiliev A. N.* // Inorganic Chemistry. 2013. V. 52. P. 1538.

406. *Zheludev A., Maslov S., Shirane G., Sasago Y., Koide N., Uchinokura K.* // Physical Review Letters. 1997. V. 78. P. 4857.

407. *Tanaka Y., Tanaka H., Ono T., Oosawa A., Morishita K., Iio K., Kato T., Aruga Katori H., Bartashevich MI, Goto T.* // Journal of the Physical Society of Japan . 2001. V. 70. P. 3068.

408. *Yakubovich O., Kiriukhina G., Dimitrova O., Volkov A., Golovanov A., Volkova O., Zvereva E., Baidya S., Saha-Dasgupta T., Vasiliev A.* // Dalton Transaction. 2013. V. 42. P. 14718.

409. *Freedman D. E., Chisnell R., McQueen T. M., Lee Y. S., Payen C., Nocera D. G.* // Chemical Communications. 2012. V. 48. P. 64.

410. *Sanz F., Parada C., Rojo J. M., Ruíz-Valero C.* // Chemistry of Materials. 2001. V. 13. P. 1334.

411. *Daidouh A., Martinez J. L., Pico C. Veiga M.L* // Journal of Solid State Chemistry. 1999. V. 144. P. 169.

412. *Sanz F., Parada C., Rojo J. M., Ruíz-Valero C.* // Chemistry of Materials. 1999. V. 11. 10. P. 2673.

413. *Danilovich I.* et al. (2017).

Index